CONSTRUCTION OPERATIONS MANAGEMENT

Tony Baxendale

THE CHANDOS SERIES ON CONSTRUCTION & FACILITIES

CONSTRUCTION OPERATIONS MANAGEMENT

Tony Baxendale
University of the West of England

CP
Chandos Publishing (Oxford) Limited

ISBN 1 902375 21 1

Chandos Publishing (Oxford) Limited
Chandos House
5 & 6 Steadys Lane
Stanton Harcourt
Oxford OX8 1RL
England

Tel: 01865 884447 Fax: 01865 884448

Printed in England

Contents

Preface vii

Acknowledgements viii

The Author ix

1 **Management of information systems** **1**

Introduction 1

Data flow 2

Structured data analysis 4

Structured system development cycle 4

Interactive management information systems 6

Initiation of a MIS project 7

Structured system development cycle approach 8

Selection of objectives 8

Summary 10

Case study 11

2 **Decision making methodology for methods of production** **19**

Introduction 19

Methodology 20

Information requirements 21

	Policies on resources	22
	Criteria influencing activity sequence	23
	Selection of plant	24
	Selection of temporary works	25
	Evaluating alternative methods of production	26
	Method statements	28
	Case study	30
3	**Construction planning**	**34**
	Introduction	34
	Tender planning	35
	Contract planning	36
	Short term planning	37
	Summary	38
	Construction planning techniques	39
	Case study	45
4	**Operational productivity**	**47**
	Introduction	47
	Work breakdown structure	48
	Work package	48
	Integration	49
	Work breakdown structure criteria	50
	Operational times	51
	Work sampling	52
	Rules for taking samples	53
	Case study	54
5	**Operational monitoring and control**	**59**
	Project progress control	59
	Integration of time and cost	61

	Alternative methods of integration	62
	Level of detail for integration	64
	Criteria for project control	64
	Computer systems for project control	66
	Case study	69
6	**Resource supply and control**	**76**
	Site cost control systems	76
	Subcontractor (or specialist contractor) control	80
	Operation execution systems	82
	Case study	83
7	**Coordinated project information**	**90**
	Introduction	90
	Concept models of buildings	91
	Coordinated information systems	92
	Activity based schedules for control	97
	Case study	99
8	**Modelling operations**	**103**
	Simulation	103
	Simulation application	107
	Case study one	109
	Detailed modelling	112
	Case study two	113
Appendix	**Master lists for classification of programme activities**	**116**
	Elements	117
	Work section	122

Preface

The aim of this book is to develop a deeper understanding of the role of the producer of building and civil engineering work in the development of the built environment. Information systems for carrying out projects, construction strategies within a decision making framework and data collection methods for production work are reviewed. Planning methods are evaluated in relation to work breakdown structure and performance requirements identified for monitoring and control that includes subcontractors and suppliers. The coordination of project data is reviewed together with the preparation and limitations of modelling construction work. The content will broaden the knowledge of all those engaged in construction work.

Acknowledgements

I should like to thank my university colleagues who have contributed to this book in many subtle ways through the research and team teaching activities that we carry out.

My family cannot go without a special thanks for the support and impetus given over the years.

The Author

Dr Tony Baxendale is a university lecturer with expertise in change and innovation in operational management issues and strategies within the construction industry. Research interests include management information systems, resource modelling and production planning. Experience has been gained working in the construction industry and since joining academia by applied research based on industrial collaboration.

Readers are invited to contact the author at the:
Faculty of the Built Environment
University of the West of England
Coldharbour Lane
Frenchay
Bristol BS16 1QY
e-mail Tony.Baxendale@uwe.ac.uk

CHAPTER 1

Management of Information Systems

Introduction

The general principles of recording data flow within a construction organisation are considered in order to specify an integrated management information system. Recommendations for the initiation of a management information system are also made together with a range of possible objectives. Structured systems development is suggested to build a logical model that shows the interrelationship of data processing. The need for an interactive system is developed together with the requirement of a structured approach based on a selection of objectives. Finally a system for main contractor control and for directly employed resource control is proposed.

Data Flow

Data flow analysis is a well established methodology and has been used for management information flow in construction organisations. Other exponents of the data flow concept have modelled the regional office and site based information systems of a large contractor and an operational integrated data management system for a house building organisation.

Systems analysis develops understanding of the flow of data through a system. Data flow diagrams are produced by analysing the progressive transformation of data throughout the system. Each transformation represents a process which is shown as a circle in the data flow diagram and lines joining the circles represent the flow of data between processes. Data flow diagrams (DFDs) are made up from four symbols as shown in Figure 1.1 together with an illustrative example. These four symbols enable a picture of the system to be drawn without commitment as to how it will be implemented. A data dictionary containing a concise definition of each element is constructed as data are identified. When the system has been totally described in terms of data, data transformation processes shown in the data flow diagram are organised into a top-down hierarchical control diagram. Finally the functions performed in each process are described using structured English to aid program coding.

An output orientated design approach has the advantage of producing minimal systems in which the precise scope of a system is defined together with the correct maintenance of data. Everything that appears as output has to be entered into the system at some point as input or has to be generated from some input. By working backwards from the outputs the minimum database and inputs for a new system can be determined. The final systems must produce answers that agree with the real world and any data that is required to be accurate must be frequently used. A feedback mechanism must ensure that data is constantly changing to reflect the updated position.

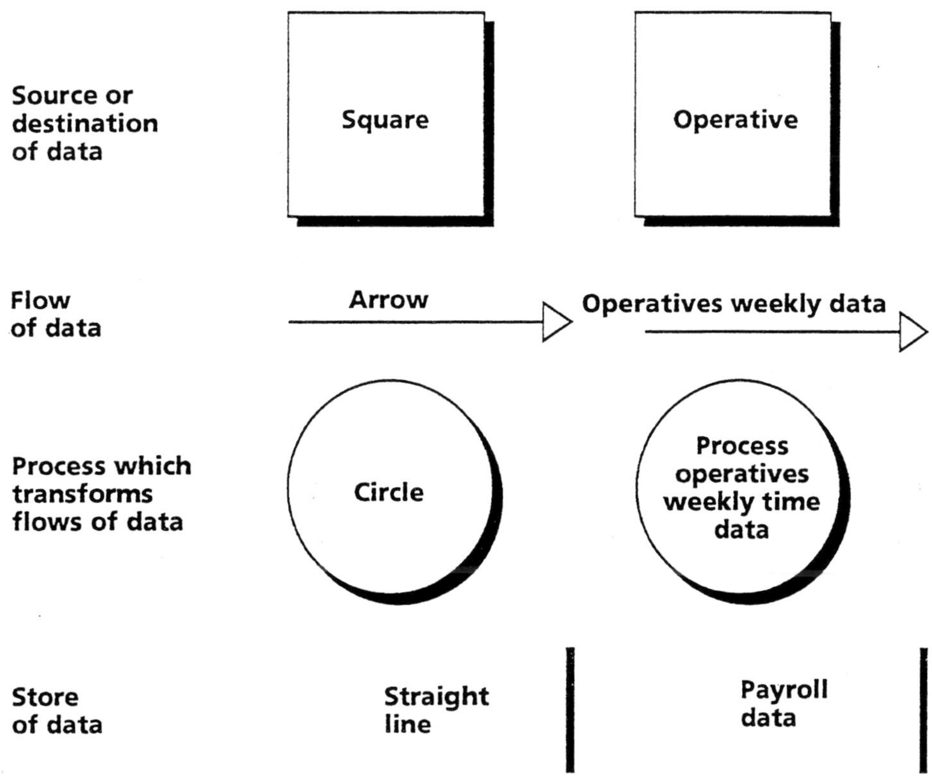

Figure 1.1 Data flow diagram symbols and illustrative examples.

Structured data analysis

The data flow diagram shows a system in terms of its component pieces with all the interfaces between components indicated. Large systems are broken down into sub-systems and sub-systems in turn are broken down for diagrammatic illustration giving a levelled set of data flow diagrams. The top diagram in any levelled set is a context diagram showing the domain of each system in terms of inputs and outputs. A context diagram for a design and build organisation information system is shown in Figure 1.2. It is at this stage that the organisation's monitoring and control procedures can be checked and inconsistencies identified and corrected. This structured approach enables the model to be broken down into sub-systems and precise procedures. Areas of commonality and divergence can then be identified and analysed.

Structured System Development Cycle

Systems analysis, design and implementation centres on building a logical model of the system and its interrelationships. Large systems have to be divided into smaller ones and details organised into an understandable structure by using a top-down approach, that is, employing hierarchical decomposition. In addition, system specifications have to be developed and maintained especially for the database and the outputs. The methodology to be employed uses features developed by Edward Yourdon in Managing the Structured Techniques, Yourdon Press, New York, 1986, which combines the data-flow, output-orientated and top-down methodologies. These features are then used in a structured system development cycle (SSDC) approach.

Structured methodology uses a series of tools for analysis and design of specifications. The data flow diagram (DFD) focuses on the logical flow of data rather than the physical flow of control. To understand the system each data flow, process and data store has to be defined

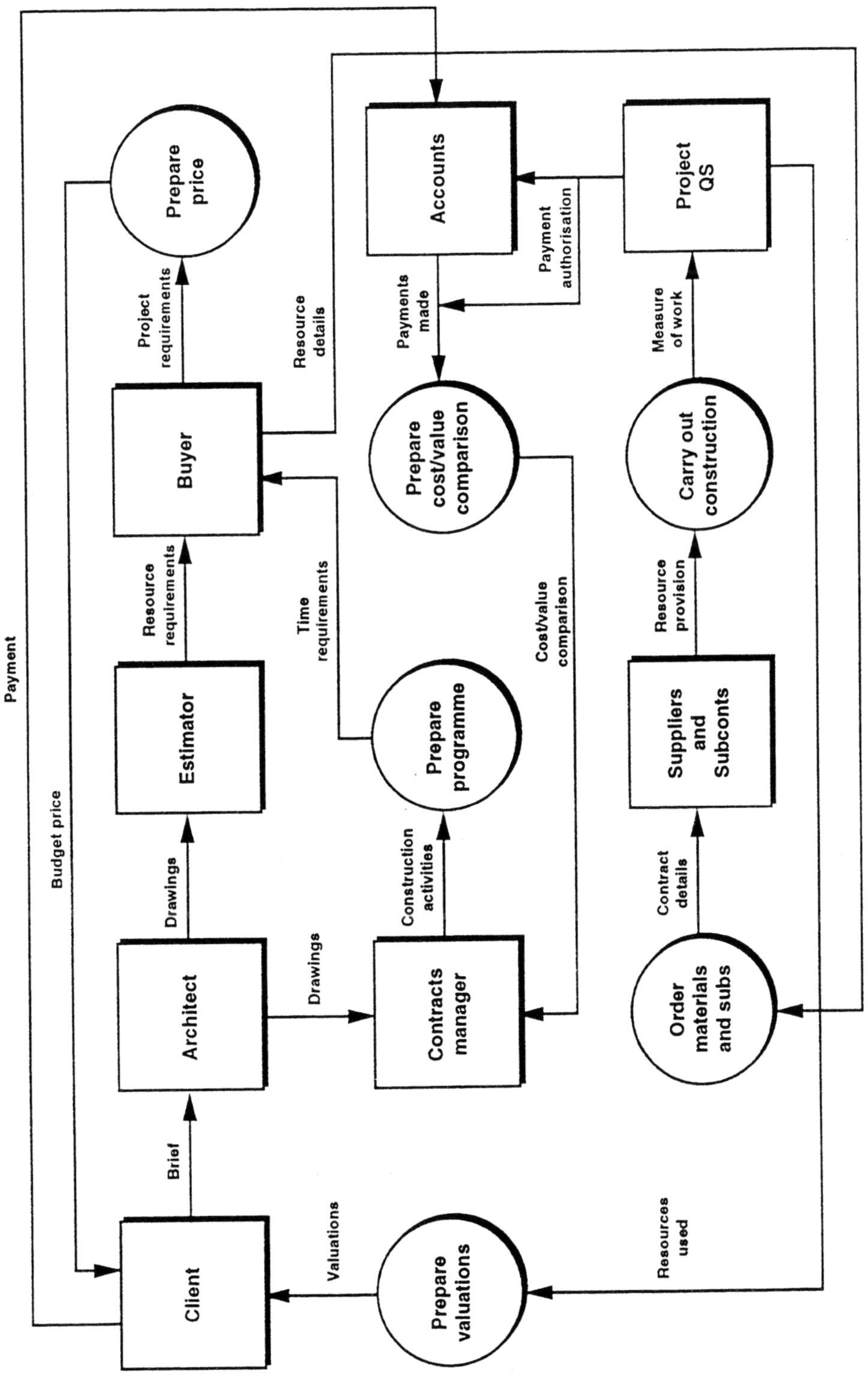

Figure 1.2 Context (level 1) data flow diagram for a design and build organisation.

by a data dictionary (DD). The data elements, structures and stores are described in detail including the process specification for each elementary process. The DD defines the rules by which data inputs are transformed into outputs. It is also recommended that structured English is used based on a restricted vocabulary that uses action orientated verbs, omits subjects and uses terms that are defined in the DD.

The objective of the structured process is to specify a management information system so that all involved have a clear understanding. The structured specifications which result should consist of the following:

- Objectives, organisation, scope, and schedule of new management information system.
- Outputs requirements (approved by users).
- Database and data dictionary.
- Inputs required to produce outputs and to store on the database.
- Processing by methods, procedures, and data communications to relate outputs to inputs as described by a levelled set of data flow diagrams.
- Internal controls to satisfy security arrangements.

Analysis will enable the flaws in currently used production control systems to be investigated with a view to improving systems in use.

Interactive Management Information Systems

An interactive management information system (MIS) centres its output on information reports and control reports. The basis of an interactive system is a common database that captures originating data as near to the source as possible to be processed to service predetermined forms of report. Information is therefore only entered into the system once and can then serve all processing requirements. Adequate and correct information can then be presented in a co-ordinated rather than segmented manner in time for management to control activities. Control reports should serve all the necessary business functions whilst monitoring actual results via feedback in order to determine whether organisational activities are proceeding in accordance with plans and standards. Clearly data processing by computer facilitates such systems.

An interactive MIS has been defined as an integrated collection of systems that are interconnected in a manner which fulfils the information requirements necessary to plan, organise, direct and control. Timely information should be produced by the system to support management and benefits claimed for such a system are:

- More effective use of employees and facilities.
- Little need to recreate data.
- Shows deviant conditions for investigating.
- Improved decision making.

The system should be capable of sending information to the user within the operational time span in which it is required to make a decision or take action. The data must be relevant to the users' needs in terms of the activity under control. The MIS should compare favourably with alternative methods when including the benefits mentioned above.

Initiation of a MIS Project

It is recognised that there are many potential problems that can hinder the development of a MIS project and guidelines that require developing to ensure a successful project are as follows:

- Involve top management from an overall standpoint in the planning phase as well as in the succeeding phase of the project.
- Provide adequate presystem planning that focuses on getting started on a project.
- Involve appropriate functional managers and their support personnel in the detailed aspects of a project.

- Obtain the assistance of the database administrator to facilitate the planning of MIS data as well as operational data.
- Recruit qualified MIS personnel from within or from outside the organisation to undertake the project.
- Provide the capability to extend the management information system into functional areas.

Structured System Development Cycle Approach

The analysis, design, and implementation of a new MIS requires a carefully structured and analytic approach. Issues have to be decided within a MIS project of WHAT must be done, HOW to do it, WHEN to do it and WHO will do it. The stages of a SSDC approach are shown in Table 1.1 with the intention that the feasibility study takes 50% of the time and the system implementation the other 50%.

Selection of Objectives

The formulation of objectives to define the scope of a MIS project is a necessary part of the investigation to provide a framework in which to operate and the constraints within which the project must function. Possible objectives are included in Table 1.2 and can centre around cost savings or emphasise faster availability of information for management decisions. Any new MIS should meet as many objectives as possible while at the same time reducing organisation costs; however, data processing costs will inevitably tend to increase.

Table 1.1 Modules of a management information system project using the SSDC approach

1. FEASIBILITY STUDY. Determines the appropriateness of making a systems change.

1.1 Introductory investigation.
1.2 Structured systems analysis.
1.3 Structured systems design.
1.4 Equipment selection.

2. SYSTEM IMPLEMENTATION. Provides the user with a fully operational system.

2.1 Training of personnel.
2.2 Structured programming and compiling.
2.3 Structured walk throughs and testing.
2.4 Parallel operations for testing the new systems.
2.5 Final conversion to new system.

Table 1. 2 Possible objectives for a MIS project

Improved information to meet the planning and controlling needs for daily operations.
Increased flow of data for meaningful information.
Increased net income from operations.
More timely information for management decision making.
Increased efficiency in organisational operations.
Elimination of conflict and overlapping within the organisation.
Improved internal control (accounting).
Better information needs to meet short-term and long-term planning needs.
Uniform and accurate handling of data with a minimum of human intervention.
Reduction in data processing costs.
Reduction in data recording and manipulation errors.
Improved client service and relations.
Increased managerial development and efficiency.
Reduction in collection of duplicate data.
Efficient utilisation of personnel and equipment.
Improved operations through greater utilisation of mathematical models.

Summary

Structured data processing involves a multi-level approach with increased responsiveness to the users' needs at the local level. Managers at local levels have more involvement with and control over the data and the system. Data processing operations that are best carried out on site stay at that level with summary information being forwarded to higher levels of management where appropriate. Low cost computing power can be placed at the points of data entry and linked with a centralised computer via a distributed communications network.

Decision support systems allow the decision maker to interact with the computer, for example by obtaining answers to a series of "what if?" questions. This approach involves the manager in the decision making process by allowing a reaction to the output, rather than accepting just one answer. The use of mathematical models as found in operations research will assist the evaluation of alternatives to structured problems as well as being helpful with semi-structured and unstructured problems.

Case Study

A System for Main Contractor Control

On the basis of investigations carried out on accounting procedures, cost accounts, valuations and cost value reconciliations, a system is now proposed. Such a system that fulfils the requirements for time and cost integration will have the following features:

(i) Activity schedule.
The activities from the contract programme form the basis of measured or builder's work, with the activities reflecting a work breakdown structure. The extensive use of subcontracting in the industry will result in each work package being associated with a subcontractor, which in turn becomes a cost centre for overall control, and each activity may be coded.

(ii) Value of work completed.
An internal valuation has a list of programme activities acting as cost centres against which the value of work completed is calculated in relation to a measure and percentage of work completed at the time of valuation.

(iii) Cost of work completed.
Subcontractor liabilities have an account code (unique to each individual subcontractor), a

contract code (unique to each project) and a programme activity code. This information is produced as part of normal invoice processing by accounts.

(iv) Cost value reconciliation.

A contractor may wish to remain with a standard format for cost value reconciliation statements where a variety of contractual arrangements are used by the organisation. If direct labour or labour only subcontractors are being used, they can be incorporated into the system by their programme activity.

No major changes are recommended to the format of the overall monthly control systems that are currently being used. But the integration of time and cost can be achieved by making the programme activity the basis of the overall monthly control system, with a coding system that facilitates the transfer of data.

Data flow diagrams are shown to include the required design features of an integrated system. The Estimating stage is shown in Figure 1.3. The work breakdown structure (WBS) forms a database for quantities and work packages, replacing the bill of quantities and the estimating buying schedule. There is now a common store of data in the WBS which does not have to be reworked. The Precontract stage is shown in Figure 1.4. The WBS data store is broken down into work packages to form the project programme. A price check on the detailed design is included which was a criticism of existing systems. The Contract stage is shown in Figure 1.5. A store of progress data has also been included in relation to the project programme and based on a measure of work completed.

This proposed system has been developed to meet the objective of devising models based on data flow diagrams to represent the practice of project planning, monitoring and control used when operating under procurement methods (design and build) which allow the contractor to prepare data.

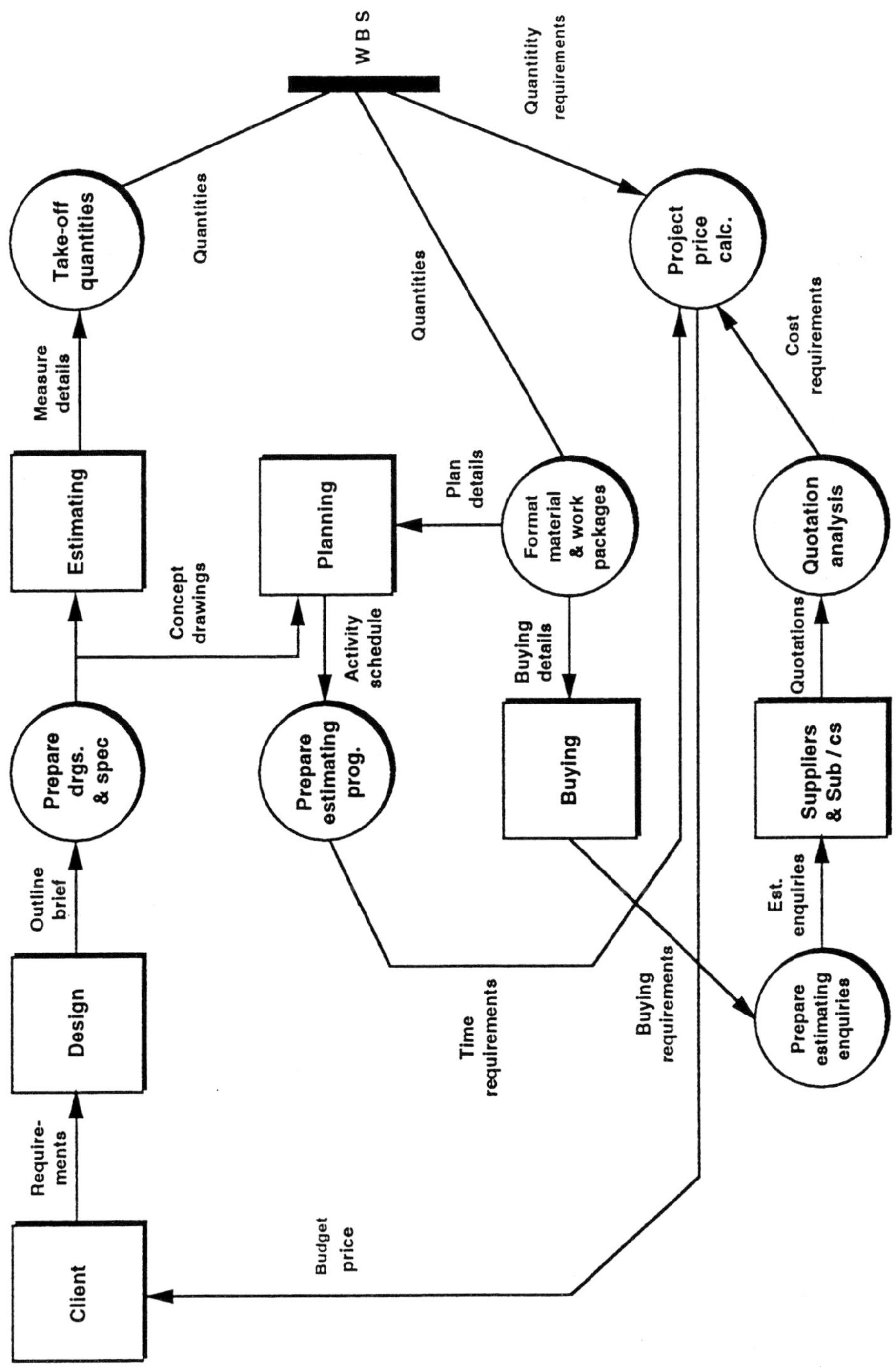

Figure 1.3 Estimating stage data flow diagram.

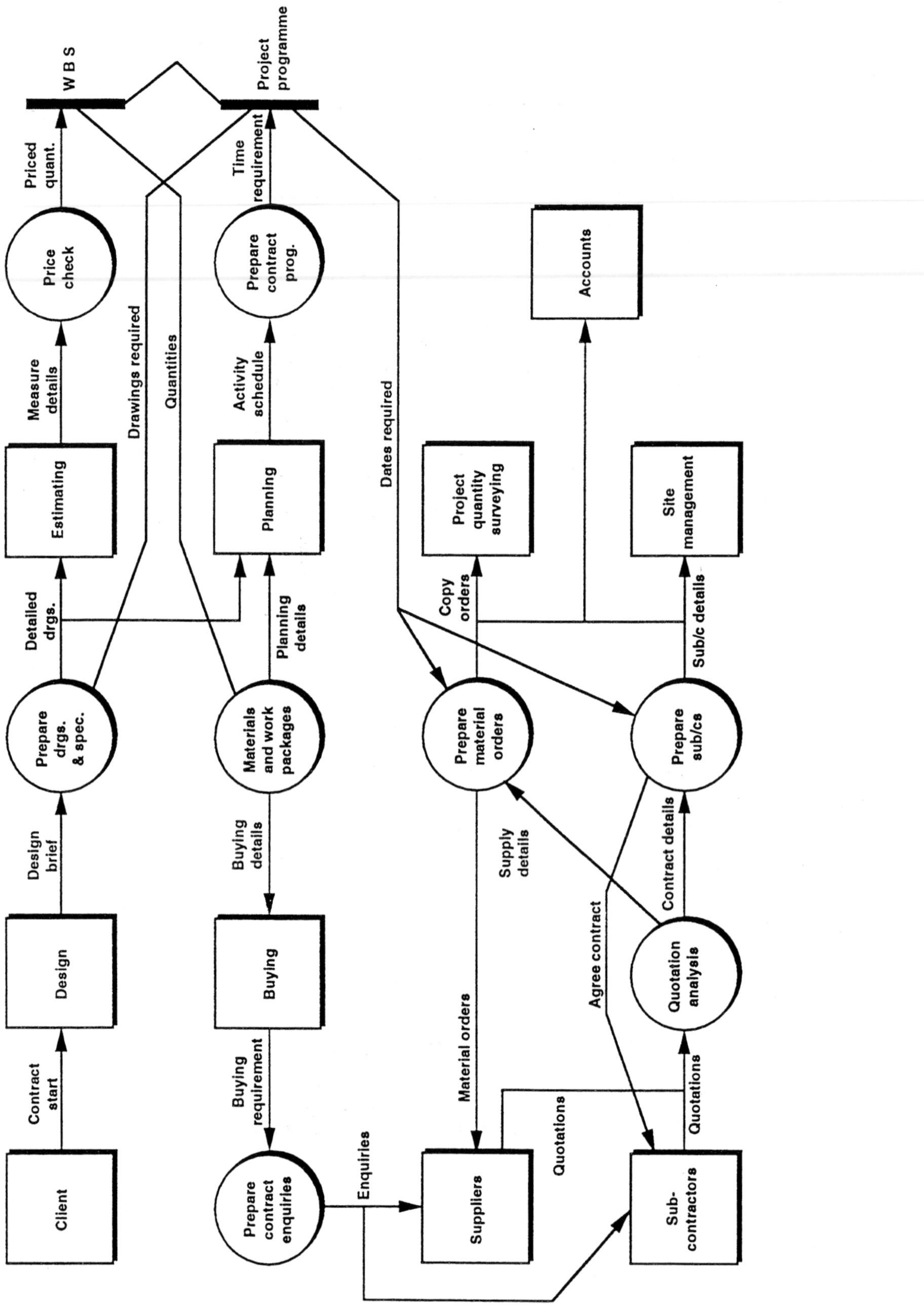

Figure 1.4 Precontract stage data flow diagram.

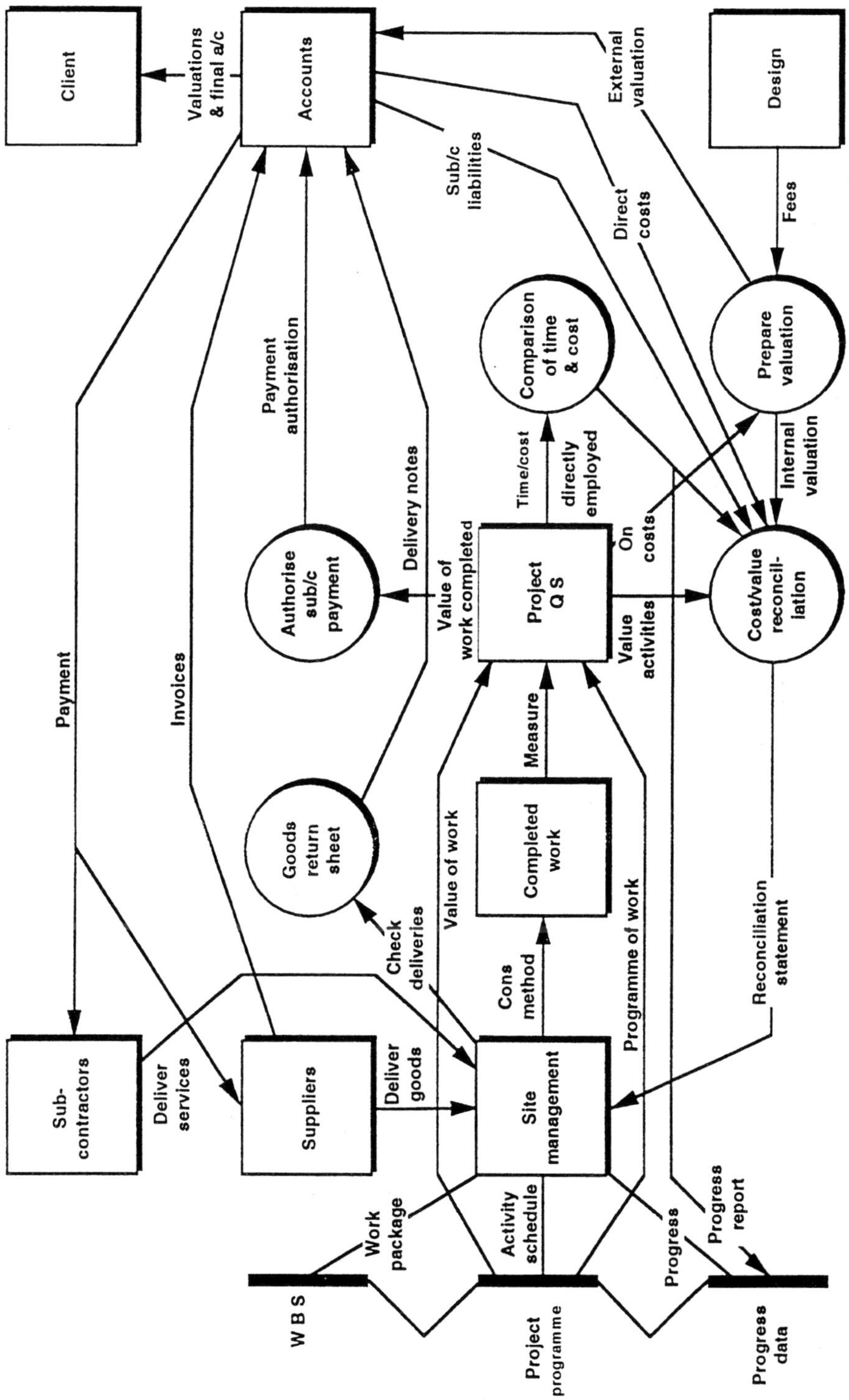

Figure 1.5 Contract stage data flow diagram.

A System for Directly Employed Resource Control

A system for the control of labour and plant has been developed and produces an integrated time and cost control system which is applicable to the main contractor for the control of directly employed labour (and plant) and to the subcontractor, who will also directly employ labour.

A proposed system that supports the criteria for project control includes:

(i) A weekly programme showing planned operations, their sequence (bar chart) and planned resource usage in terms of operative or plant hours. A progress percentage is added at the end of the week for each operation.

(ii) A weekly allocation showing actual hours spent by each operative or machine on each operation for each day together with a summary total.

(iii) A weekly reconciliation showing planned value and planned cost from the programme based on an hourly rate, actual value from the percentage progress and actual cost from the allocation sheet; the planned and actual gain or loss is then calculated for each operation and totalled for the week.

Operations in a weekly time and cost control system will be based on, for example, a labour-only subcontract or a trade gang and may only involve one activity or cost centre. Each operation can be coded if required by the originator of the programme using the coding system outlined in relation to work packages. The main variation in coding between operations will be for type of resource and location.

A database is also recommended to be part of the same system to provide feedback. The preferred unit for data will be hours, as cost will relate to a particular period of time and allowances will have to be made for inflation. The database will include:

- Contract name, number and classification of type of work.
- Operation and code in relation to work package.
- Measure of work completed in relation to an appropriate unit.
- Time allocated to the operation for labour and/or plant.
- Output in hours per unit for feedback.

A collection of outputs for similar operations can be made across all projects or for a particular classification such as industrial development. The original data should be maintained for inspection and any rogue outputs examined for validity. An averaging of outputs may be applicable or a frequency distribution will be appropriate for some operational research techniques.

A data flow diagram representing the system is shown in Figure 1.6. This will be a second level DFD and relates back to the project quantity surveyor and the comparison of time and cost processes shown on Figure 1.5, a first level DFD. The contract data will be based on the project work breakdown structure. This proposed system has again been developed to meet the objective of devising models based on data flow diagrams to represent the practice of project planning, monitoring and control used when operating under procurement methods (design and build) which allow the contractor to prepare data.

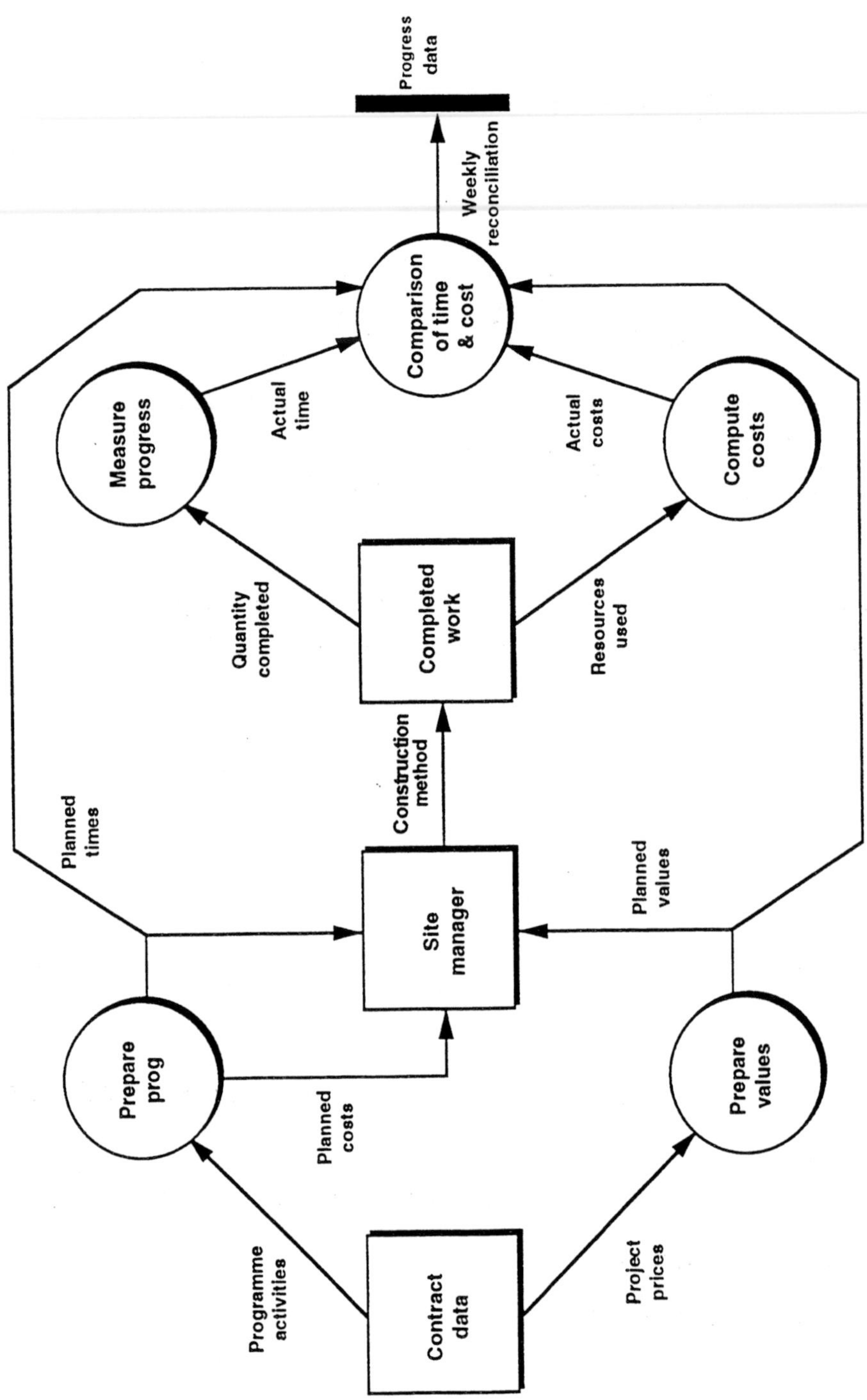

Figure 1.6 Directly employed resource control (level 2).

CHAPTER 2

Decision Making Methodology for Methods of Production

Introduction

The aims of this methodology for decision making are to identify production opportunities and problems and also to develop strategies for the production of complex contracts. Decisions will be both of a 'policy' and a 'hard' decision in nature. Decisions will need to be made upon:

- Level, type, balancing of mechanisation to manpower resources.
- Extent and type of temporary works.
- Specific safety and protection measures.
- Storage facilities.
- Sequencing and timing of activities.
- Direction of working and access requirements.
- Whether or not the production method is economical and achieves the performance

and quality requirements of the building.

The decision making process is **not** linear and requires constant reappraisal; be prepared to choose solutions, test them and make changes.

Methodology

Information

Consider what information you require to make a decision. On most large contracts, the information provided may well be substantial. You will need to sort out what is relevant, assess its quality and identify whether there are any information gaps.

Analysis

Analyse the information, identifying any significant production problems and opportunities. Normally the number of main problems will be small, say less than five.

Activities

Prepare a list of activities and decide on the level of planning you wish to enter into.

Resources

Consider the overall policies on resources for the contract as a whole, for example, the level of mechanisation.

Sequence

Consider the sequence of building: there are always alternatives, for example a sequential sequence compared with a concurrent sequence. Identify clearly the interrelationship between activities. When developing policies and 'hard' solutions assess the **risks** involved, particularly potential damage to property and injuries to people.

Alternatives

Consider alternative solutions for each activity on which decisions need to be made. Solutions must satisfy cost, time, health and safety constraints, and quality standards specified by the designer.

Evaluation

Make an evaluation for the alternative methods of production for each activity.

Decide

Make provisional decisions.

Review

Test ideas for building as a whole or in stages, for resource balancing, cost and overall time. Be prepared to make changes or try again, for example change the sequence or method, before making a final decision.

Communicate

Finally, communicate your ideas using method statements where appropriate, schedules of resource requirements, a site layout diagram and programmes of work.

Information Requirements

Information can be obtained from the following sources:

- Architectural and Consultants Drawings.
- Bills of Quantities:

  appreciate overall dimensions of the building, its height, shape, basement depth;

  appreciate the nature of the structure, how structural stability is achieved;

  ascertain storey-heights.
- Identify the main elements, their materials and components; ascertain their quantity, weight, position in the building; establish maximum weight to be lifted.
- Ascertain the extent of wet as opposed to dry construction.
- Ascertain degree of standardisation and repetition in design.
- Confirm the nominated suppliers and specialist subcontractors; appreciate relationship of services to the building structure.

- Specifications

  appreciate the performance standards to be achieved;
  identify dimensional tolerances to be achieved for each element, and critical parts of the construction where tight tolerances have been set;
  identify standards of finish;
  check limitations on pour sizes, sequencing of pours, for example avoiding trapped bays, and allow curing periods;
  ascertain quality control standards and tests required.

- Site Visit

  check position, type, condition and age of adjacent buildings;
  check proximity and dimensions of boundaries, trespass possibilities, position of major roads, potential surcharge pressures on excavated areas;
  assess environmental conditions, that is potential noise and vibration restrictions;
  ascertain access and egress points;
  appreciate site space and degree of congestion, possible restricted plant movement;
  appreciate site topography.

- Soil Investigation Report

  appreciate the nature of the soil, its ability to support itself, soil hardness;
  check levels of ground water.

- Contractual Conditions

  ascertain the overall contract period and any phased handovers.

Policies on Resources

Consider the objectives for the contract: is the intention to produce the building to predetermined quality standards at the optimum cost or to complete the contract within a shorter period of time? In other words, what is the priority between time and cost? The balance of plant and labour has to be considered, that is plant intensive as opposed to labour intensive solutions. This will be influenced by resource availability, the nature of construction, site constraints, resource capabilities, operative satisfaction, productivity requirements and comparative output costs (see case study at the end of the chapter). Material storage alternatives have to be considered: materials can be stored on or off site, or fixed into position when delivered. The allocation of space and use of central compounds or progressive storage within the building will be required. Finally any potential conflicts in policy have to be resolved.

Criteria Influencing Activity Sequence

Examples of production criteria that influence the sequence of activities:

- Devise a sequence which will reduce contract time and accelerate cash-flow; in particular the intention will be to make the building watertight as early as possible to enable services and finishing trades to proceed.
- Organise work in packages for optimum resource usage.
- Provide continuity of work for resources where possible.
- Develop safe sequences especially in congested areas, that is separate out activities and avoiding working overhead.
- Provide access for following trades without causing unnecessary obstructions.
- Provide space for material storage or access for plant.

Examples of performance and quality criteria that influence the sequence of activities:

- Ensure structural stability during construction, for example against wind forces
- Develop sequences which minimise potential damage by following trades and which provide protection to materials and processes affected by inclement weather.

When devising the sequence it is advisable to assess the construction in relation to drawing details and assess any potential work-fragmentation problems.

Selection of Plant

It is not just an item of plant that is being selected, but possibly a combination, for example excavators and lorries, hauling and placing concrete. In other words, you are selecting a system which, in addition, has to be correctly balanced with human resources. Labour is expensive and therefore optimum gang sizes should be planned. Failure to do so will increase costs as over-manning cannot produce as efficiently. While many items of work have established optimum gangs, for example bricklaying, with some operations the gang size cannot be constant, mainly because the degree of operational difficulty varies. Consider the differences in placing concrete in a raft foundation compared with a waffle floor: clearly the concrete has to be placed and vibrated with more care and possibly finished differently in the waffle floor. To achieve the same output more operatives will be needed. The labour/plant relationship and correct manning must be fully appreciated and even if the work is sub-contracted, such knowledge is still important if the efficiency of the subcontractors is to be evaluated and controlled.

The main factors influencing selection are likely to be the shape and construction of the structure, its relationship to ground conditions and contours, and the relative position of boundaries including any adjacent buildings. Some significant points to consider are as follows:

- Quantities of materials to be handled may vary on each storey (often at roof level). It may therefore be more economical to use a different method of handling than that used for the other storeys.
- Tower cranes are not just for high buildings, their wide coverage (especially if rail-mounted) may also be suitable for sites with blocks of low level buildings, in close proximity, with limited access around them.
- Sloping sites can pose additional problems: handling materials by wheeled machines may be difficult and dangerous. Materials may need to be lifted down by cranes standing on levelled ground or concrete may need to be pumped.

- Materials delivered to site may be packaged and the weight of unit loads may vary. What is convenient to manufacture may be neither convenient nor economic to site handling. Remember that plant selection is only one item in a chain of events: manufacture, supply, delivery, site unloading and storage, onward handling to the final location. In other words, consider compatibility when selecting plant for handling materials.

The selection of plant also depends upon knowledge of its capabilities and limitations, for example whether it has linear, two or three dimensional movement; the maximum load which can be lifted and at what radius; its reach whether it be a crane, pump or excavator; its ability to manoeuvre in restricted spaces and move in poor ground conditions; its output. A machine's mode of operation has to be appreciated; for example, tractor shovels need to move across areas when excavating, whilst 360° backactors are stationary. It is necessary to be aware of the attachments and accessories available to realise versatility and the correct manning levels have to be known.

Having established the operating conditions and matched up machine capabilities, decide on alternative plant systems bearing in mind availability, then evaluate them in terms of production and economic implications.

Selection of Temporary Works

The decision for formwork and falsework includes the type of system, that is preformed proprietary as opposed to traditional made in place. The criteria to consider include: speed and ease of assembly and dismantle, adaptability and versatility, skill level, other resource requirements and hire as opposed to buy costs. The choice of formwork materials will have to be considered as will ease of handling and protecting, maintenance levels and costs, re-usage levels, surface finish requirements, ability to meet tolerances and weight. The height of wall panels is often critical and is influenced by mechanisation available for lifting, congestion of reinforcement and likely segregation of concrete, function of wall and implications of construction joints, implications of hydrostatic pressure related to member sizes and their costs. The number of forms to make or hire will depend on the number of re-uses required and the economic implications, the time constraints of the contact, the curing periods, weather and striking times, the amount of labour on site and the need to maintain work continuity, and the cost of forming stop-ends.

Resources required to install an earthwork support system will include materials, plant (cranage and driving equipment) and labour, as well as time and cost implications. The extent and choice of these are influenced mainly by earth, hydrostatic and surcharge pressure, depth, length to be retained, and whether the system should be watertight or not. The soil hardness influences the selection of pile driving equipment and the ability of soil to support itself. The risks to property and people, site congestion influencing lateral support arrangements, and the noise and vibration levels to be tolerated are further considerations. Whether the system is to be recovered or not is important as temporary works are expensive. Methods should be explored as to how the permanent works can be used to reduce temporary work involvement.

Evaluating Alternative Methods of Production

Alternative methods should always be considered and the final arbiter of choice will be on the basis of meeting building performance and quality standards, production requirements and cost. When evaluating plant and method selection the production implications should include consideration of the following:

- Any significant safety regulations which would affect the proposals being suggested.
- The production capabilities of each proposal, that is the time to complete an activity, and whether or not the plant system has the retrieval capacity to make up time if the contract falls behind schedule.
- Whether each proposal will be utilised efficiently, is any idle time likely to occur, and the implications for a contract when major items of plant are not in use.

- The site planning required to maximise the full potential of a major item of plant, for example a fork-lift truck will require good access (use of permanent or temporary roads), space around storage areas/delivery points to enable ease of handling, and loading out points designed to take account of the load carrying capacity of the machines. The position of the crane in relation to the storage/delivery area is particularly critical to avoid unnecessary travelling of the jib and a reduction in output. The site must be planned around the activities of the machine for the most efficient results.

When evaluating cost it must be realised that the above points regarding production all have cost implications. Two different situations are likely to arise: (a) where handling equipment is required to perform a mixture of roles, for example lifting and placing a variety of construction materials; and (b) where only one material is handled on a productive output basis, for example, insitu concrete. In (a) direct comparative costing has to be done in relative cost per week terms. In (b) more detail is possible and costs related to output can be established. Before such cost comparisons are attempted, it is crucial that competing methods are compared on precisely equal terms. For example, if ready-mixed concrete is involved in one method and site-mixed in another, the cost of materials should also be included.

Three Stage Approach

To be complete, the cost analysis must be carried out as a three stage operation as is shown in the case study.

(i) Make a direct comparison of alternatives, taking into account installation, dismantling, any temporary works used, transport charges, running and maintenance factors, gang sizes, safety precautions, hire rates, production output. The direct comparison must include the **whole** process, for example mix, haul, and place for concreting.

(ii) Assess possible intangible factors, for example the effect of inclement weather. In other words, assess risks.

(iii) Assess the influence that method selection will have on related items and the contract as a whole, for example a large pour decision might affect formwork costs.

Points to consider on making a decision are that achieving quality in the construction is of paramount importance, otherwise the method has failed and extra costs for correcting the work will be incurred. The decision has to be in line with policy objectives for the contract. Consider which has priority with regard to production time and optimum cost. Always assess potential risks and the adverse possibilities of your decision. Always comply with safety regulations. Consider the suitability and availability of plant and equipment already owned by your company. If the plant is only partly suitable, consider the implications of using it compared with the cost of hiring from an outside company a machine considered to be more appropriate. Successful plant selection will ultimately depend upon efficient site control and minimising the effects of adverse problems.

Method Statements

An extract from a method statement used to communicate production decisions in shown in Figure 2.1. Method statements are used for repetitive activities, where careful analysis may lead to substantial savings being made due to an efficient work method being employed. They are also useful in congested situations where there are possible fears of overcrowding leading to low output. They illustrate major activities of size, for example bulk excavation work, new technical operations not familiar to site staff, and complex activities in general. Problems encountered in their use and which have to be guarded against are adequate information not being available and the time and effort required to produce them. Also site staff may not be briefed on their content and the design may change during construction.

Activity:

Excavate to 20.500 level.

Method:

Set out work and set up site controls.

Commence excavation using a tracked excavator and 7 tonne tipper lorries.

Access for lorries to be via the back lane.

Therefore dig to commence at High Street side, working towards the rear of the site.

Support excavation at front and rear of site by leaving a batter at this stage of the excavation.

Part of the batter to be extended to provide access ramp for lorries to next stage of excavation.

Protect the edge of the excavation.

No ground water is expected during the summer.

Resources:

1 No. excavator

5 No. lorries

Quantity:

750 m^3

Output:

15 m^3/hour

Time:

1.5 weeks

Figure 2.1 Method statement example.

Case study

This case study emphasises the basis of decision making for alternative resource solutions.

Large Raft Foundation to Office Block

This is a classic example of the large concrete pour situation. Until the raft is complete little other work is possible and so such situations lend themselves to consideration of high rate/large volume pouring as potentially economic situations. The case study illustrates a comparative cost analysis and also emphasises why a total cost has to be considered.

The situation to be analysed is that of a concrete raft 1.4 m thick with a content of 2207 m^3. The tender method statement allowed for site mixing with the concrete being placed by two tower cranes with skips. The reasoning behind this decision was to include continuity of plant, as the output from the site mixer was adequate for economic production of concrete for the structure above. Two tower cranes were also required for the superstructure for concreting and other handling work. The estimated time for the raft construction was 10 weeks.

At the pre-construction stage it was decided to re-analyse the raft to see if economies could be made and the time for completion speeded up. It was established, by agreement with the structural engineer, that large volume pours could be used with ready mixed concrete. Three pours were possible at volumes of 900 m^3, 262 m^3 and 1045 m^3. Only the smallest pour could be completed by the ready mixed concrete supplier during normal working hours. The other two pours would have to be carried out at the weekend which would involve site operatives being paid overtime rates.

A detailed cost study is now shown to assess the viability of the alternative ready mixed pour using concrete pumps against the original approach using site mixing and cranes.

Method and Cost Analysis ONE

Site mixer + 2 no. tower cranes

Direct comparison items only included			Cost £/m³
Mixing			
Plant cost (all inclusive)			
and Labour mixing		09.20	
Materials		48.80	
Waterproofing		20.80	
		78.80	78.80
Placing and distribution			
Crane hire (all inclusive)			
Tower crane (1) 4 weeks earlier	8040		
Tower crane (2) 14 weeks earlier	43740		
Less 20 t crane for 7.5 weeks	11800		
	39980 divide by 2207		18.10
Labour			
1 no. operative 50 hr x 4 wks x £5.10 =		1020	
1 no. operative 50 hr x 6.5 wks x £5.10 =		1657	
4 no. operatives 50 hr x 8 wks x £5.10 =		2040	
		4717 divide by 2207	02.14
Effect on related items:			
Steel fixing - similar in both cases			
Formwork (stop ends) 450 m			
Material cost	4880		
Labour cost	25310		
	30190 divide by 2207		13.80
		Total cost £ per m³	112.84

Method and Cost Analysis TWO

Ready mixed concrete + 3 no. pumps

Direct comparison items only included Cost £/m³

Ready mixed				
Materials	46.08			
Waterproofing	20.80			
Pumping additive	01.20			
	68.08			68.00
Placing and distribution				
Pump hire: £350 per day (all inclusive)				
3 pumps x 3 days x £350 = 3150 divide by 2207				01.43
Labour				
Pour 1	11 operatives x 12.5 hr x £5.10	=	703.80	
	Overtime	=	126.23	
Pour 2	10 operatives x 8 hr x £5.01	=	408.00	
Pour 3	16 operatives x 11 hr x £5.10	=	897.60	
	Overtime	=	122.40	
			2258.03	
		divide by 2207		01.02
Effect on related items				
Steel fixing - similar in both cases				
Formwork (stop ends) 60 m				
Material cost		650		
Labour cost		3380		
		4030 divide by 2207		01.80
	Total cost £ per m³			72.33

To complete the overall analysis the effect on the contract as a whole needs to be considered as well as any intangible factors. The original method estimated a time of 10 weeks for the raft completion, whereas the time possible with ready mixed concrete and pumping is 7.5 weeks, thereby saving 2.5 weeks on the contract time. This will result in a saving on contract preliminaries and overheads, estimated as £18750.

There are some intangible factors that may be taken into account. With ready mixed concrete and pumps there would be considerable problems if there were any plant breakdowns and standby plant may have to be made available at an additional cost. The site mixed concrete set up would require storage areas as a precaution against delivery failure. Finally both methods are vulnerable to weather conditions and the time of year has to be taken into account.

There are further technical considerations when making large concrete pours:

- Potential cracking; extra quality control required.
- Good co-ordination required between main contractor, subcontractors and suppliers.
- Good access is required to ensure short pumping distances, facilities for washing out pipe lines, support to pipe runs and walking platforms over reinforcement.
- Extra labour is required to allow for gangs to be relieved as pouring will be continuous; additional resources may be a problem.
- Re-usage of formwork may be reduced.
- Following operations will have to be brought on to site sooner if the work is speeded up.
- Continuity of work for operatives and plant not involved with the concreting operation may be a problem.

However, on the basis of the cost comparison the choice of ready mixed concrete and pumps appears to be the most economic approach to the foundation under consideration.

CHAPTER 3

Construction Planning

Introduction

An organised system is required in the planning and control of construction projects. The general objectives of an information system to aid management are as follows:

- Provide an organised and efficient means of measuring, collecting, verifying and quantifying data reflecting the progress and status of operations on the project with respect to programme, cost, resources, procurement and quality.
- Provide standards against which to measure or compare progress and status. Examples of standards include network programmes, control budgets, procurement schedules, quality control specifications and working drawings.
- Provide an organised, accurate and efficient means of converting the data from the operations into information. The information should be realistic and recognise:

  the means of processing the information (manual or computer),
  the skills available, and
  the value of the information compared with the cost of obtaining it.

- Report the correct and necessary information in a form which can best be interpreted by management and at a level most appropriate for the individuals who will be using the information.
- Get this information to the correct managers and supervisors, that is those in a position to make best use of it.
- The information must be received in time so that, if necessary, corrective action can be taken on those operations that generated the data in the first place.
- In keeping with the principles of management by exception, the following objectives may be added:

  identify and isolate the most important and critical information for a given situation, and

  give it to the right person as quickly as possible for their consideration and action.

The key components of feedback control include means of measuring and controlling processes, information processing, effective reporting and taking corrective action.

Tender Planning

The planning considerations at the tender (or pretender) stage should be thorough and exhaustive, even if the ultimate result only serves only to substantiate the initial impression that the job is not an attractive one. While the production of an estimate remains the responsibility of the contractor's estimator, sound planning for construction can make a substantial contribution towards the production of a realistic tender. The extent to which planning can be applied should be related to the complexities, or otherwise, of the form of construction and the planning resources available to the contractor.

Information Required at the Estimating Stage

The information that can be produced at the estimating stage is as follows:

- Site and locality investigation report.
- Report on visits to architects' and engineers' offices.
- Programme for the job.
- Labour schedule.
- Plant schedule, mechanical and non-mechanical.
- Schedule of site supervision.
- Method statement, including the design of temporary works.
- Site layout showing access, offices, stores, services, plant and material storage areas.
- Schedule of basic quantities and anticipated production rates.
- Schedule of nominated and own subcontractors and suppliers.

Programme of Work

The length of time required for the major operations should be shown and the degree of detail expressed on the programme has to be related to the complexity of the proposed works. Apart from enabling the time periods for site supervision, mechanical plant, scaffolding, temporary buildings, reuse of formwork and the like to be established, the programme will also substantiate or disprove the overall time period allowed for the job.

Contract Planning

The main purposes of contract (or precontract) planning are to produce a plan which will:

- Enable the work to be completed within the contract period.
- Show durations which are consistent with carrying out the work as profitably as possible.
- Enable progress to be conveniently shown and so form a method of control.
- Communicate the plan in a simple manner to all concerned.

The plan is the method of tackling the project. This is the project manager's concern with the assistance of site staff who will be directly in charge of the work. They will know from experience how to set about a new project and what labour and plant are available. A specialist planning engineer can also assist in the preparation of the plan. Planning is a combination of method study, experience and common sense. It is the planning engineer who can help in translating the plan into a programme and so commits the plan to paper.

The programme technique used must be suitable for the type of work and there are many types of work and a number of different planning methods. There is no reason why a combination of techniques should not be used to suit the various aspects of the project. The programme format must also suit the company's methods and personnel.

Short Term Planning

This is possibly more important than contract planning and deals with labour and plant, looking usually four weeks ahead with the first week considered in greater detail. It should be done on the basis of labour and plant costs in the estimate for the job as far as this is possible. Clearly each company requires to devise its own system as site based costing will be involved. Short term planning also has to be within the framework of the contract plan and gives:

- Detailed control within the contract plan and within the value of the labour and plant estimate.
- Detailed planning that should not be done at a distance. Circumstances change and there may be unforseen problems such as ground conditions, weather, availability of resources and so on.
- Involvement of the site staff and operatives who carry out the work. The short term planning should be partly their decision and include their knowledge and experience.
- Opportunity to leave short term planning to site management.

Summary

Programming techniques rather than planning techniques have been considered. Planning is dealing with how, in what order, when and with what resources the work will be done. This may include detailed method studies and the assistance of others from within the contract team.

Some basic principles that can be applied to planning are as follows:

- Know the construction details.
- Develop a basic plan.
- Try to make major decisions as soon as possible, eg. method of materials handling.
- Do not neglect such points as:

    doing the deepest work first,
    working towards the access to the site,
    tackling certain jobs at an appropriate time of year.

- Develop a more detailed contract plan and programme.
- Prepare schedules from the programme.
- See that all programmes and methods are communicated to those concerned.

- See that the programmes are implemented and used for control.

Finally, whatever planning and programming methods are adopted, keep the contract programmes as simple as possible and use short term planning on the construction site to deal with the detail. Do not forget the importance of communication, implementation and control.

Construction Planning Techniques

There are many different types of work in the construction industry that range from an individual low rise building to a complex development of multi-storey buildings. No one planning technique suits them all. The most commonly used techniques are now reviewed.

Bar (or Gantt) Chart

Bar charts are used for normal, comparatively simple programmes. They are ideal for short term planning and outline tender planning. Bar charts are simple and easy to prepare and understand. They can be as good as more sophisticated methods if care is taken. Progress can be shown by drawing a progress bar underneath and adding the percentage completed on a daily or weekly basis. There is room for giving information on items of work, in particular adding dates when information is required. The drawbacks are that bar charts do not show dependencies between operations, nor those operations which are critical. The consequential effect of delays is not shown and complicated jobs cannot be dealt with adequately. Omissions are also quite common because of the overall nature of the programme. An outline bar chart is shown in Figure 3.1.

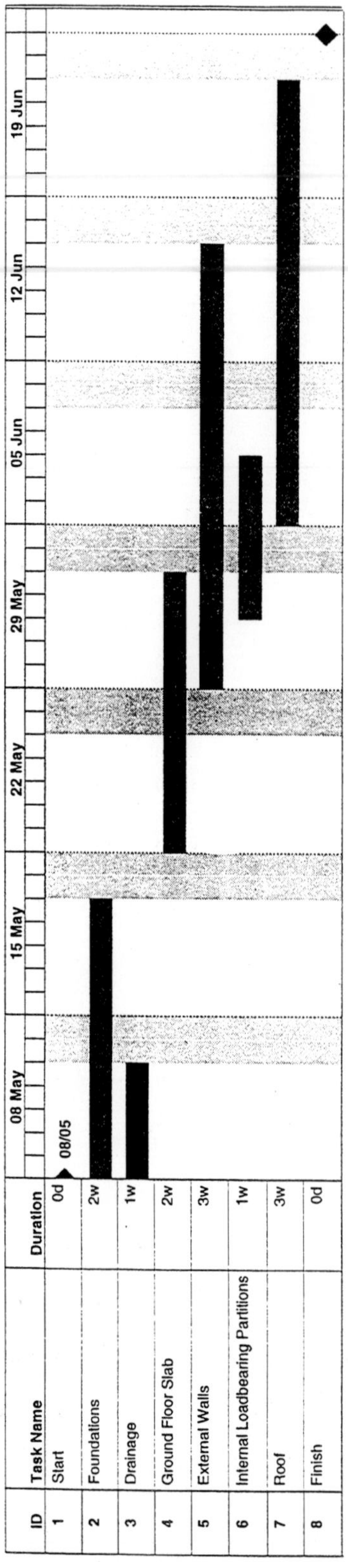

Figure 3.1 Outline bar chart.

Network Analysis

Network analysis or critical path methods (CPM) are used where a precise planning tool is required and for any job where the dependencies between operations are important. They are also used for all jobs which are too complicated to be planned by any other method.

The procedure for preparing a network is as follows:

- Planning — Decide on the activities in a project.
  Put the activities into a logical sequence in the form of a network diagram, showing how activities relate to and are dependent upon each other.
- Scheduling — Allocate duration times for each activity.
  Calculate the time during which each activity can occur, earliest project completion, the critical path and the spare time or float available for any activity.
  Critically analyse the result.

The network is best drawn to a timescale for presentation, in weeks against calendar dates and showing the critical path and float time. The critical path is the longest sequential path of activities in a project which connects the beginning and end of the project and gives the shortest overall time. Finally the resource requirements (labour, plant, etc.) have to be balanced to arrive at realistic time periods within a practical project completion date. The precedence diagram (activity on node) format tends to be used rather than the arrow diagram (activity on arrow) format. This is because precedence diagrams are regarded as being easier to use, having no dummy activities, and it is easier to revise the diagram.

A simple method of resource balancing can be achieved by picking the more important resources and letting the rest balance themselves. When allocating durations to the activities keep as far as possible to very simple multiples of the resources, for example one gang or one crane. Using a timescaled network with activities at their earliest start times, resources can be shunted about within their float periods to achieve the best readily attainable balance. Detailed resource planning is best left to the short term planning that takes place on site during the project. Bar charts are the only other method to which resource balancing is applicable and a very similar technique can be followed.

The advantages of using networks are that they show activity relationships at a glance, pick out the critical activities, show float on non-critical activities and show the effect of delays on following activities. They are a good discipline for preparing an overall programme providing the procedure for planning and scheduling is followed, and minimise the possibility of omitting items. (Networks are a good method of communication when plotted to a timescale but they are poor for illustrating repetitive jobs (eg. housing) and process type jobs (eg. roadworks).)

It is recommended on large projects that an overall network is prepared with subsidiary, more detailed networks for parts or even single activities from the overall network. The overall timescale will be in weeks or months with the detailed network in weeks or days. With typical building work it is better to proceed from the general to the detailed and not the reverse. Often there is no point in going into detail in non-critical areas that can be left for short term planning. If a network analysis technique is necessary for the planning, then the dependencies between the activities are important and therefore should be shown in the method of presentation. Timescaled networks can have a long life and do not easily get out of date in terms of showing progress. Progress can be shown by shading in the time bars to indicate percentage completion or by dropping a near vertical line (or string) cutting the bars at the position reached. An example of a network is shown in Figure 3.2.

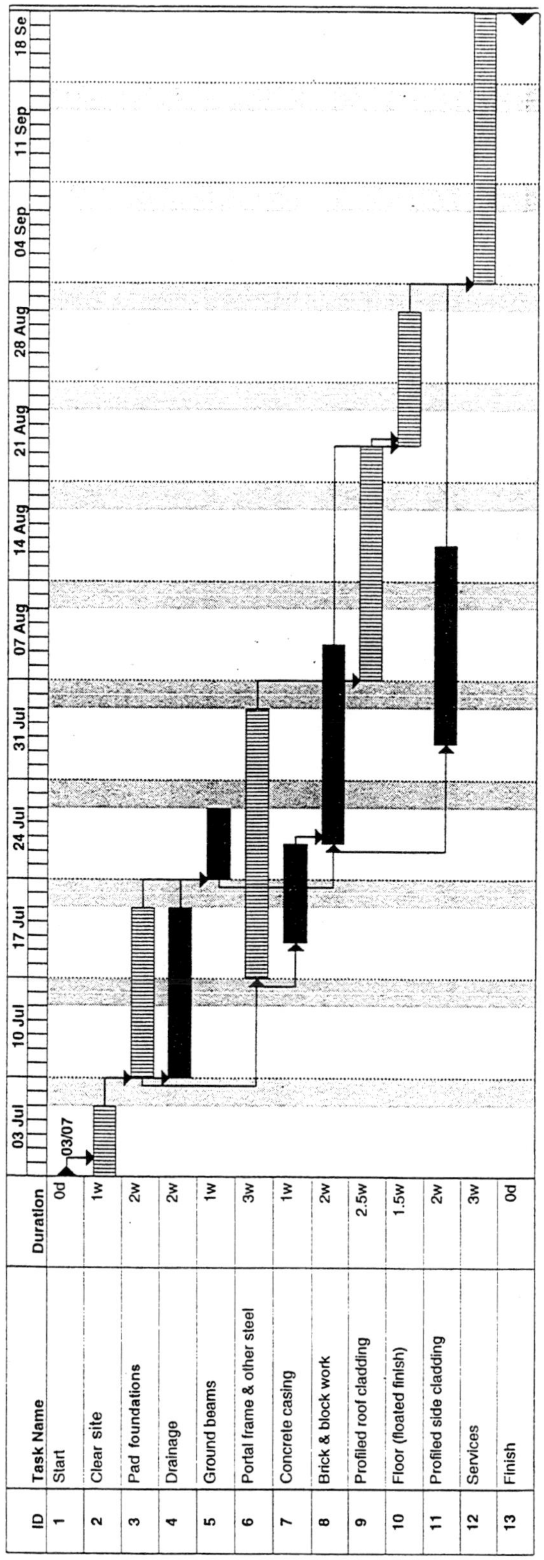

Figure 3.2 Time scaled network.

Output Graphs

Output graphs make an excellent simple programme for checking outputs for a dominant activity or material and even as a programme to work to. Output or cumulative output is plotted against time on a daily, weekly or monthly basis. Output can then be balanced across the programme for such activities as excavation, concreting and the like. Graphs are often used as a subsidiary to an overall programme with progressing by plotting the actual output for comparison with the planned output. The advantages are that a simple graph shows progress well and can illustrate how the whole job is progressing for such fundamental items as brickwork or road base and surfacing. Importantly the cumulative graph can show trends by comparing the slope of the line of the actual output with the slope of the target output.

Stage Diagrams

A series of shaded or coloured in plans show the work being done at a particular stage of a job. An area of the plan can show dates when work is planned to be carried out. Progressing is not normally attempted as stage diagrams are normally used to illustrate a detailed programme. However, the date the work was completed can be added or areas crossed off. Stage diagrams are used for illustrating trade work in confined areas by location or for multi-level work constructed in definite stages with a change of activity. Stage diagrams can explain a complicated job in a quickly understood and straightforward way. They come into their own where location is not clear by description alone and are good for checking the proximity of operations and access routes thereto. The disadvantages are that they become very cumbersome if a lot of information is to be shown and the preparation (if coloured) is very time consuming. Also a diagram can get quickly out of date if variations to the planned work occur.

Case Study

Which planning methods, or combination of methods, are recommended for the projects described?
[A] Bar chart, [B] Network analysis, [C] Output graph and [D] Stage diagram.

(i) A shopping centre with basement storage and delivery access, retail areas up to four storeys, covered pedestrian areas and external car parking with a value of £40 million.

(ii) A speculative development of warehouse distribution units, constructed from a steel frame, metal cladding and a concrete floor slab with a value of £2 million.

(iii) A nineteenth century, six storey, dockside warehouse being refurbished as open plan offices. The existing structure will remain but all finishes and services are being replaced at a value of £6 million.

(iv) A city centre flyover with four traffic lanes and constructed in insitu concrete. Traffic flow at the existing cross roads is to be maintained at all times, underground services diverted at a value of £16 million.

(v) An eighteen storey office block constructed with a steel frame and precast concrete flooring. The external cladding is of decorative brickwork with many different types of special bricks.

(vi) Building work to a factory production line during a two day close down period. Seven different subcontractors will be working in the area with a value of £0.6 million.

Solution to case study:

Method [B] is recommended for project (i).
Method [A] is recommended for project (ii).
Methods [B] and [D] are recommended for project (iii).
Methods [C] and [D] are recommended for project (iv).
Methods [B] and [C] are recommended for project (v).
Methods [B] and [D] are recommended for project (vi).

CHAPTER 4

Operational Productivity

Introduction

Productivity is defined as the quantitative relationship between what we produce and the resources which we use. Higher productivity is concerned with increasing output from the same, or smaller, use of resources. Productivity is an overall conception which can be expressed in terms of output from labour, or from services, or from the capital invested. In the long term it is only through an increase in productivity that economic success can be achieved. In the context of this chapter productivity is related to an examination of work in an operational context which leads to the investigation of all the factors which affect the efficiency and economy of the situation being reviewed, in order to effect improvement. An operations manager is mainly concerned with the control of resources on the construction site. Any cost control system tends to make comparisons with the estimate only in retrospect. What is required is a measure of the use of resources or level of productivity. A work breakdown structure is firstly defined in relation to production activity, together with relevant criteria for control models. Next, the sources of information from which operation times can be derived is considered and finally the work sampling method of investigating productivity is illustrated.

Work Breakdown Structure

The requirement for a systematic way of defining and identifying the division of production activity has been established in Chapter Three. The need is for a Project Breakdown Structure or Work Breakdown Structure (WBS) that usually involves an elemental dominance in definition such as, foundations, frame or external cladding. WBS is a hierarchical method of breaking down a large project into the services and work items required to produce the end product. The result is a hierarchy of portfolios of smaller projects, each with their responsibility defined and allocated. The concept can be applied to smaller projects of lesser complexity where the breakdown can be more obvious or simplistic. WBS descriptions which are synonymous with network activities in a master or overall programme (produced at pre-tender or pre-contract stage) create the link between elements and their progress. By allocating resources we can then proceed to cost.

Work Package

The work package classifies the varying resource requirements of operational activities. The dominating influence on classification will be work section, trade, construction form or material and location. A clearer statement of construction logic is produced for short-term or stage planning. It should aid measurement for progress reporting. The work package breaks down the WBS, for example a suspended concrete floor slab could appear under formwork, reinforcement and concrete work packages, an electrical installation subcontractor's work under location by floors. It is important, of course, that work packages are defined in a standard form for transfer to a database. Retrieval for future planning or estimating is important for the full economic utilisation of such a system.

In detailed terms there are three dimensions to be recorded in a work package:

- Physical characteristics (quality, specification and location).

- Resourcing (personnel, plant, materials, work space and time).
- Relationship to priced items of work.

A standard system of WBS will be required that has to be constantly applied to a firm's work to reconcile accounts with and between jobs. It will be necessary for the level to lie between summarised general information (too crude for meaningful comparison from job to job) and too detailed a collection of information (requiring too much administration). The standard classification of network activities will include physical characteristics, resourcing and pricing data.

Integration

A uniform and consistent system of communication is required where data input or information produced by the various control functions can be sorted and reported on a common basis. Sharing data electronically is not enough for effective integration. For example, the value of work for an item and its cost must be allocated to the same work package if comparison is to produce accurate and meaningful information. The WBS will form the basis of communication between all control functions involved with a contract and therefore the use of a common WBS is required for the effective integration of a management information system. Integration is not the collection of data but the correlation of data so that variances can be calculated in relation to a standard. In overall terms a WBS is required to be:

- So simple that it can be easily understood by all those expected to use it.
- Such that it is significant and meaningful at each level of use, both from a data collection and reporting point of view.

The WBS is at a summary level for reporting time and cost in relation to master programme activities. The work package is at a detailed level for allocating labour and plant or labour only subcontract use of resources. An example of a standardised description and coding application for an in situ reinforced concrete internal wall is shown in Figure 4.1. The coding system relates to that devised in Chapter Eight. Additional coding digits will be required for location and subcontract identifications that are contract specific.

ALL superstructure		|2|0|
Superstructure internal walls		|2|2|
ALL concrete work		|E|0|
In situ concrete		|E|1|
Formwork		|E|2|
Reinforcement		|E|3|
WBS code		|2|2|E|0|
Work package code:	Concrete	|2|2|E|1|
	Formwork	|2|2|E|2|
	Reinforcement	|2|2|E|3|

Figure 4.1 Work Breakdown Structure and work package coding.

Work Breakdown Structure Criteria

As a result of the requirements established criteria are now devised for production control models based on a WBS:

- Each work package is to be identified as a network programme activity (each activity will also have to be a work package).
- The definition of each work package for each contract will have to be identified and agreed by all who use the system (estimator, planner, contracts surveyor and site manager).

- Data will have to be collected in relation to the work contained in each activity.
- Activity data will have to include resources that are to be the subject of control (labour, plant, site overheads and materials in order of priority).
- The input of data will be site based and the output of information reported for both contract and company.
- Updating of work package content will be necessary for any variations in quantity or value.
- The coding of work packages in a standard form that is transferable from contract to contract for comparison.
- A statistical data collection facility in relation to a historical input to a database (mean, range and distribution).
- Recognition that different criteria will apply to subcontractors who are controlled on site by performance and quality only.
- Easily understood by all those who use the system.

Operational Times

In deciding how long an operation should take the sources of information available are the labour and plant content included in the bill of quantities for estimating, performance data from previous contracts, and standard and synthetic data from a work study library of information.

Labour and plant contents included in the bill of quantities provides a direct tie-up between the content of the bills and the overall programme. This makes comparison in overall terms easier to apply and will show whether the contents in the bill are accurate. This source is only useful at the preconstruction stage where the programme operations are not considered in detail and therefore average outputs are used which are like those in the bill. Many items in the bill will have to be grouped under an operation heading, eg. brickwork and all the associated extras. This can be a very laborious job. Some bill items will have to be broken down for sequence studies, eg. formwork, reinforcement and concrete in a structural frame. A point of warning is that the bill may not be accurate.

Performance data from previous projects are a good basis for operations in a short term programme and where operations used are similar to those on previous projects. They can also be used for overall programme operations as average outputs can be calculated from the data available. The same basic information may be used for incentive target schemes. It is easy to compare with work carried out on previous operations and information should be readily available from operational records. Performance data used for a combined cost control and planning system can also be used as a basis for future estimating. Feedback information of this kind may not be entirely reliable, particularly where an incentive system is in use as moving time from one operation to another can occur unless tight supervision is exercised. It also relies on the accuracy of time allocation sheets and it is not known what the level of performance was when work was carried out.

Standard data from a work study library of information is the most scientific and accurate method of assessing expected output, at a known level of performance for a prescribed method. An account can more easily be taken of site conditions and accurate information used for setting targets. The problem is that it takes some considerable time to accumulate the necessary data, it is not easily adaptable for use at the pre-construction stage and is not easily reconciled with a bill of quantities. Experience has found that the practice of work study is rarely carried out for construction work mainly because of the temporary nature of the production process.

Work Sampling

Productivity on site can be investigated by carrying out a work sampling exercise. The results from work sampling productivity studies are best used in pointing out to operations managers where work can be carried out more effectively. Data is collected on what each operative or machine is doing at small intervals of time with particular interest being taken in non-productive activities for analysis and reduction.

Work sampling is a means of measuring the effectiveness of the site in using its resources. The results of the sampling can be available at the end of the working day which will allow managers to make decisions on improving the use of their resources. Work sampling involves observing and classifying a small percentage of the total activity of a project. With a large enough sample, the level of activity of any operation can be estimated. The accuracy of the estimate will improve as the number of samples increases.

Work sampling (also called activity sampling) is used to determine the activity levels of operatives and plant. Readings are taken at intervals of time and those taken for each category will give a result close to the actual percentage.

Rules for Taking Samples

The following general rules must be observed for work sampling:

A sample should have at least 384 observations. This will give a statistically acceptable result every time.

The observations must be categorised at the instant the operation is observed. The observer should not rationalise what an operative has just completed or is about to start.

Every activity must have the same chance of being observed at any time.

Observations should have no sequential relationship. It follows from this that cyclic operations will have to be randomly observed.

The basic characteristics of the work must remain the same while observations are being carried out. This is important if comparisons are to be made between different sets of observations.

Case Study

Inefficiencies in construction operations often exist because managers are too close to the work or too busy to see them. What is required, therefore, is an approach where work can be investigated in a planned way to improve productivity. The main categories of work can be identified:

- Directly productive, which is an expenditure of time in making the building grow.
- Indirectly productive, which is an expenditure of time in preparing to make the building grow.
- Non-productive.

Work is only effective when it is directly adding to the completed product. Each observer will have to define effective work, depending on the operatives or plant being sampled. The classification of work is not rigid and may change depending on circumstances. The aim is to show where there are inefficiencies.

An example of a productivity study is given in Figure 4.2 together with a summary in Figure 4.3. Study details are entered at the top of the sheet. Every two minutes (round interval) the activity being carried out is noted against the material, for example laying pipes. A round interval can be between half and five minutes depending on the number being observed. 384 minimum observations are required, as stated previously in the rules for taking samples. Non-productive activity is also noted under the headings of waiting, personal needs,

Reference: 2	Sheet No: 1	Contract: HOUSING
Date: 16 October	Observer: AB	Operatives/Plant: 2 layers, 1 driver
Start time: 08.00	Finish time: 13.00	1 ganger (part time)
Round interval: 2 minutes		Notes: Connections to manholes

Activity	Material	
Laying and jointing	Pipes	///// ///// ///// ///// ///// ///// ///// ///// ///// ///
Setting out		///// ///// ///// ///
Mix by hand	Concrete	///// //
Mix by excavator		/////
Excavate by hand	Pipes	//
Excavate by machine		///// ///// ///// ///// ///// ///
Place by hand	Concrete	///// ///// ///// ///// ///// ///// ///// ///// ///// /////
Place by excavator		///
Excavator travelling	///// ///// ///// /////	
Excavator maintenance	///// ///// ///// /	
Resting	///	
Waiting	///// ///// ///// ///// ///// ///// ///// ///// ///// ///// ///// ///// ///// ///// ///// ///// ////	
Personal needs	///// ///// ///// /	
Communication	///// ///// ///// /////	
Walking at work place	///// ///// ///// ///// ///// ///// ///// ///// ///// ///// ///// /	
Walking to and from	///// ///// ///// ///// ///// ///// //	
Extended break	///	
Late start/ Early finish	///// /	

Figure 4.2 Work sampling sheet

WORK SAMPLING SUMMARY

Reference: Study 2 Subject: Drainlaying, 2 layers & 1 driver
Time: Overall 08.00 to 13.00 Observation interval: 2 minutes
Observations: Possible 426 Actual 428 Error 0.5%
Accuracy: Better than +/- 5% for any category

ACTIVITY	NUMBER of observations	PERCENTAGE of observations	
Lay and joint pipes	48	11	
Setting out	18	4	
Mix concrete by hand	7	1	
Mix concrete by excavator	5	1	
Excavating by hand	2	-	
Excavating by machine	28	6	
Place concrete by hand	50	12	
Place concrete by excavator	3	1	
Place m/hole rings by hand	11	3	
			39%
Excavator travelling	20	5	
Excavator maintenance	16	4	
			9%
Resting	3	1	
Waiting	84	20	
Personal needs	16	4	
Communication	20	5	
Walking at place of work	56	13	
Walking to and from	32	7	
Extended break	3	1	
Late start	6	1	
			52%
	428	100	

Figure 4.3 Work sampling summary sheet.

communication, walking at place of work, walking to and from place of work, extended break, late start and early finish to the working day. The observations are totalled and shown as a percentage for easy comparison.

The breaking down of work into small categories enables the assumption to be made that operatives will work at a standard rate. It is the not working categories that are of interest but the analysis of the non-productive activities. A check is also made between the number of possible and actual observations with any allowable error being less than two percent.

The case study shows a directly productive time of 39%, an indirectly productive time of 9% and a non-productive time of 52%. It is generally considered that productive activity of less than 60% is unsatisfactory and will require further investigation. Every operation will have to contain a certain amount of relaxation and contingency time, but this should be not more than 40%. The main point arising from the study is the 20% waiting time which indicates that the excavator is not working efficiently. The excavator driver was also only excavating with the machine for 6% of the time.

In general the main reasons found for low productivity are:

- Walking to and from the place of work when materials are stored at a distance from the work position.

- Waiting due to inadequate supervision, non-delivery of materials or restricted access making transportation of materials difficult.

- Overmanning in relation to work available or working area.

- Imbalance of gang in terms of operatives or machines.

- Poor time keeping at the start and finish of the day and extended breaks during the working day.

- High personal needs content, mainly socialising with others on site.

- Bad weather preventing work.

Productivity studies are straightforward to apply and can be carried out by junior managers or a visiting specialist. The studies are questioning the site organisation and not the operatives' work input. A directly productive time of 40% and below is often found on construction sites and the aim of supervision must be to improve total productive time to 70% and above. As the case study shows it is the non-productive time and its reduction that requires to be analysed. The derivation of any general principle that can be applied to other operations is of limited value as the circumstances prevailing on any particular site may be unique to that site. The aim, therefore, of any study will be to improve productivity on that particular construction site.

CHAPTER 5

Operational Monitoring and Control

Project Progress Control

A project is dynamic and must respond to changing conditions if it is to be completed successfully. There is a continual need for reassessment and reappraisal of the project plan. Factors affecting an existing plan will include:

- Changes in the technical specification.
- Changes in the required dates.
- Changes in relative priorities.
- Revision of activity duration estimates.
- Reassessment of resource requirements for individual activities.
- Changes in resource availabilities.
- Inaccuracies in planned sequences.
- Technical difficulties.
- Failure of deliveries.
- Unexpected weather conditions.

It is therefore necessary to have a monitoring system which generates feedback that enables corrective action to be taken. There are usually some deviations that do not allow the project to proceed in accordance with the plan. It is therefore necessary to review operations periodically and to update or replan when a change is revealed. Close or detailed control of resources is not always considered. Close control is where resources are fully considered during the initial stages of the project and the timing of every activity is fixed, so as to obtain optimum use of resources. Flexible or overall control is often exercised during the initial scheduling of the project. Flexible control only considers resources to avoid peaks in key resources or those resources that are in limited supply. The frequency of review will depend on the overall duration of the project and the timescale on which the activities are measured. In general a weekly programme should be reviewed weekly and a daily programme daily.

If an accurate picture of progress is to be achieved, a systematic method of collecting data must be used. There is also a need for speed, so that there is as little delay as possible between reporting and the taking of corrective action. The two basic requirements of a project action reporting system are that every activity should be exactly as defined on the programme and that the progress information at the report date is as up to date as possible. A project action report gives such information as:

- Activities that should have started.
- Activities that are due to start.
- Activities that should have finished.
- Activities that are due to finish.

A visual form of monitoring can be used in addition to the project action reports when the programme is drawn to a timescale. A progress line is drawn starting at the review date and passing through each activity depending on the percentage complete. The progress line will also show trends. The project may be ahead of schedule but the progress showing a trend to return to schedule. This is the time to take action to prevent the project getting behind schedule. If progress is shown in this way on a timescale programme, minor changes in schedule duration can be catered for and the programme only has to be redrawn for major changes in duration or dependency alterations.

The feedback of project progress at regular intervals as the work proceeds translates estimated dates into actual dates. Therefore the proportion of the programme representing jobs still to be done can be reevaluated to identify new values of float or even a new critical path. If delays fall within the float time and resource availability limits, rescheduling is possible. If, however, the affected activities are critical, or there are resource problems, 'crash' action may have to be taken. The forms of crash action include:

- Adding resources.
- Transferring resources.
- Changing the logic of the network.
- Substitution of equipment, methods, and the like.

Integration of Time and Cost

The problem of integrating time control and cost control is now analysed by reviewing the functions of planning, budgeting and control systems. Aspects of direct and indirect integration are considered together with the level of detail and the use of control indices. A case study of an integrated system in practice is used to test the monitoring of time and cost.

The integration of a planning (time control) system and a budgeting (cost control) system gives rise to a number of problems. However cost information cannot be meaningful unless it is related to a time frame.

The major functions of a planning system are to:

- establish a work programme (or schedule of objectives),
- report progress of the work,
- identify deviations from plan,

- show a projected schedule for the remaining work,
- show the effect of contract changes.

The primary functions of a budgeting system are to:

- establish cost objectives,
- report historical costs to date,
- show estimated costs to date,
- compare estimated with actual costs,
- show the effect of changes.

The functions of control systems are:

- to set standards (performance objectives),
- to measure output and compare it with standards,
- to react to unacceptable deviations by signalling them,
- to respond to signals by providing a basis for decision making in respect of any corrective action,
- to facilitate that action and
- to review standards.

Alternative Methods of Integration

An information system that integrates time and cost has two important requirements:

- Integrated data collection based on consistent cut-off dates and single data sources.
- Reporting that reflects both cost and time effects on project performance trends.

Two alternative approaches are now reviewed:

The Direct Alternative

Project costs can be broken down to agree with scheduled activities and will directly eliminate the structure mismatch problem. At the planning stage costs must be budgeted by activities and during project execution budget accounts are opened and closed as activities are started and finished. Thus performance problems are detected early as each activity reports progress or is completed. This approach of work packaging is considered to have serious limitations. The practical problem of collecting actual cost by activity is that scheduled activity breakdown often changes during project life and is more volatile than cost codes. Scheduling can require activities to be defined at levels of detail where cost data collection can be cumbersome. The reverse is also possible, for example tracking bulk material costs by activity. The factors that drive costs and schedules are not all the same. A material may not be purchased for only one activity. Operative and plant resources may be used across more than one activity.

The Indirect Alternative

Cost and schedule breakdown remain different but both the structure and processing of data are interfaced where appropriate. There will be a common dependency on resource links for both cost and schedule. Reporting costs in the same format as schedules is suggested as being an artificial way of suggesting that integration exists. Progress is reported from the scheduling system which in turn integrates with a resource control system. Finally the resource control system is integrated with cost reporting, financial and accounting systems in which data is summarised into information. The indirect approach can be attractive to contractors who manage construction resources in the field. Implementing this approach requires a careful use of coding structures and information systems interfaces. The emphasis is on achieving cost/schedule integration without compromising the individual objectives of each control function.

Level of Detail for Integration

A view is that it is impossible to integrate cost and schedule systems effectively at summary level. Measurement of the criticality of the work being performed is required in order to give proper weighting to reports. To be successful cost/schedule integration must be performed at the detailed level with interrelation of the systems requiring:

- the development of the cost system and the schedule system at the level of detail each independently requires,
- the determination of which cost items and which schedule items can be correlated,
- the transmission of meaningful and required data between the systems,
- the realisation that it is not possible to determine a direct correspondence for transmitting all information,
- the analysis of the information being generated by direct contact with the personnel involved.

Cost and schedule systems are compatible but not identical; not every item can be integrated. Correlating data and analysing the resulting reports will improve the ability to develop, monitor and control both cost and schedule. The system used in Figure 5.1 later in the chapter has been developed to meet the objective of devising a data coding system, based on a work breakdown structure for integrating time and cost control, that meets the requirements for coordinating building project information.

Criteria for Project Control

Control of time and cost can be integrated through work packages. These can be defined in a standard form for continuity of feedback. The objectives and functions of each contributing resource data system should remain unaffected. A work package will relate to a programme activity and to an estimate of cost for that activity. It is the presentation of processed data in the form of variances in time and cost that should assist managers in controlling projects. The control system should also afford strategic aims against which the satisfying of user needs by the system can be measured.

It can be reasoned that a model to represent the practice of project planning, monitoring and control will have the following objectives:

- Model the projected work in terms of planned progress and cost using a common database.
- Permit a comparison of actual progress and cost against planned.
- Identify the sources of variance and quantify them in terms of time and cost.
- Provide a means to explore operational decision making by allowing independent access at varying levels of detail.
- Record data on resource usage in relation to performance and cost to facilitate future planning and estimating (historical records).
- Relate the flow of funds for the project to the policy and performance of the organisation as a whole (accounting records).
- Recognise that user requirements are of paramount importance and that information from the system must be capable of influencing the decision maker.

The possibility of integrating time and cost data for site based control, at both the overall and detailed level, has been established. An integrated system has been devised in the case study that overcomes the practical problems of producing control variance and forecasting reports. The case study has tested the hypothesis that it is feasible to output data from a construction site that monitors and integrates the functions of both planning and job cost accounting, and shows it to be possible.

There is a need to integrate time and cost control systems, but for such information to be effective it must also be practical. Automated computer project control systems are available, but their success is more a function of how they are used rather than what the software can do. An alternative view is that the controlling of past performance and the processing of future projections should be separate, as the incorporation of the time factor in cost control complicates the procedure unnecessarily.

Indirect integration can be achieved by using a common coding system which allows all existing systems to interface information. Careful use of work breakdown structures for programme activities at the project level in relation to subcontracting will overcome the problems of direct integration, provided that cost data is prepared in relation to the activity format. A MIS should have the ability to focus on various aspects, such as cost or activities, as there may be differences in emphasis required by the site or by the firm.

Computer Systems for Project Control

There are many project management systems available which have cost modelling capabilities to a varying degree. There are ready-made standard packages that can be used after one or two days familiarisation and system languages that allow a wide range of features to be built in to give more research flexibility but require programming.

There is a view that bespoke software may be advantageous. An example of a model of the way in which information technology can be managed in a medium sized company now follows. The system looks after accounting, estimating, planning and cost reporting for the group and also links senior managers in a network using electronic mail. The company started in the traditional area of accounting with proprietary software but were soon tailoring the programs to meet their own needs. Most of the system is personal computer (PC) based but there is a growing network which cannot be supported on a PC basis.

One of the reasons for using their own software was the need to share information throughout the organisation as most people use the system in their everyday work. Contract information is available and it is a question of assembling it to meet the needs of individual managers. A manager is encouraged to provide an outline of requirements and then a data processing department tries to emulate them within the system, without the use of manuals. A project history is developed for each contract from the initial contact with the client, through the estimating and tendering stage to site work and final completion.

There is still resistance to the system from some members of the organisation but priority has now been given to surveyors, contracts managers and directors as they have been without computer assistance. It is reported that contract surveyors now feel more in control having got away from a paper based system. The valuation is entered in directly and is fed automatically to the production of a cost report and the subcontractor payment system. The cost report will be available to senior management at the same time who are able to check progress via their own terminals and passwords.

The case for site managers to have a PC with a modem link to head office mainframe is often argued, where access to, for example, network planning expertise can be facilitated. Once the work has been finalised, the site should be able to investigate revisions and prepare progress reports using their own PCs and software. The control of documentation on site is often a problem and the setting up of a computerised document retrieval system can help to solve it. The aim will be to rationalise procedures and resources used for the acquisition, condensation, manipulation, retrieval and presentation of information in support of the site manager's daily functions.

A project management system example is now given that uses proprietary and home-made programs which are partially integrated. Cost accounting is run on a mini computer and handles accounts and issues work instructions. Automated cost allocation and task completion reports form the basis of interim payments to subcontractors. There is a link to PC based estimating and budgeting packages as well as a document control system. Cost management is carried out on a spreadsheet program and is linked to cost budgeting and cost planning. A proprietary project planning system is used for stand alone application and to link to the central planning team. A word processing package is used for running on networked PCs and document control (issue and approval) is tracked using a database package. Traffic management also has a program to control material requirements for each subcontractor on each building. It is interesting to note that payment is made for an agreed percentage completion against specific work descriptions, without detailed measurement. The accounts are then automatically processed, an attribute that is assisted by the coding of all trades and all the activities within each trade. Costs can therefore be accurately allocated to each building location and provide a link of information back to the estimators.

A view expressed is that many Project Management Systems can be categorised as: critical path method based acting as tools for project planning and control; estimating and quantity surveying systems that do not relate to construction activities; accounting systems that are tied to commercial considerations. These limited approaches will not support the information management required on large projects. An information management hierarchy for reporting time, cost, quality and feasibility is as follows:

- Performance indicators to measure trends.
- Management exception reporting of activities that do not match defined criteria.
- Action reports that are sorted for individual responsibility.
- Registers of information that will mainly be accounting reports.
- A database that links all available project information that can be searched, sorted and reported.

There are currently, therefore, many approaches to the application of computer systems to project control:

- Standard software packages.
- Bespoke software.
- A mixture of both standard packages and bespoke software.
- Valuation orientated software which also produces a schedule of activities.
- Site based terminals linking to a main computer by modem.
- On site PCs with data transfer by disk.

All approaches aim to integrate management information and provide feedback. Large complex projects can require all of these attributes for project control using a database management system.

Case Study

The application of an integrated time and cost control system is considered for a project. An integrated reporting system (IRS) is developed, with the emphasis on cost, to enhance an adequate existing short-term programming system. The reporting system is based on work packages which are broken down in relation to resource usage and location. A variance report is also calculated for each work package as a basis of forecasting for the total project. An application of the system in use is shown in Figures 5.1 to 5.5 and the problems found with data collection and presentation together with the implications for site management found are also reviewed.

The project ran without any major problems as regards progress. However, the failure of the programme to define distinct work package activities created problems. This was highlighted where whole sections of detailed work were missing. On this project no builder's work appeared in the budgets or on the programme, and labour was released from the site too soon; this led to daywork bills for two operatives for two months. A simple mistake in grouping builder's work with service subcontract work could easily have been avoided had the IRS system been used.

With the traditional system the invoicing of materials provides the cost control point. However, this can lead to problems when assessing the cost of individual activities. This becomes more pronounced at the project level. At one stage of the project the profit to date was shown as £20,000 less than it should have been. This was due to profits made on materials which did not come to light until after all the invoices had been corrected due to lack of information.

Project	*Industrial Development*		Project code	*4 2 3 9*
Week No.	*18*		Date	*18 May*
Work package / Activity	*Edge beam to floor slab*		Activity code	*E 1 1 3*
Location	*Workshop / Canteen*		Location code	*2*
Resource	*Labour*		Resource code	*1*

Type	Description	Quantity	Rate	Total
Ganger	*Bank machine, lay dpm, set up road forms (30m),*	*15 hrs*	*3.00/hr*	*45.00*
Labourer (a)	*pour concrete and vibrate, trowelled edge (60m),*	*15 hrs*	*2.50/hr*	*37.50*
Labourer (b)	*power float (45m²), protect (45m²).*	*8 hrs*	*2.50/hr*	*20.00*
Labourer (c)		*8 hrs*	*2.50/hr*	*20.00*
Youth Lab.		*4 hrs*	*1.75/hr*	*7.00*
		Total cost £		*129.50*

Figure 5.1 Work package report for labour

Project	*Industrial Development*		Project code	*4 2 3 9*
Week No.	*18*		Date	*18 May*
Work package / Activity *Edge beam to floor slab*			Activity code	*E 1 1 3*
Location	*Workshop / Canteen*		Location code	*2*
Resource	*Plant*		Resource code	*2*

Type	Description	Quantity	Rate	Total
Excavator	*Grade formation*	*4 hrs*	*15.00/hr*	*60.00*
3t dumper	*Transport concrete*	*4 hrs*	*7.50/hr*	*30.00*
Poker vibrator	*Placing concrete*	*4 hrs*	*0.50/hr*	*2.00*
Power float	*Finishing concrete*	*10 hrs*	*1.00/hr*	*10.00*
	Note: Long hire road forms and small tools in general site plant			
		Total cost £		*102.00*

Figure 5.2 Work package report for plant

Project	*Industrial Development*		Project code	*4 2 3 9*
Week No.	*18*		Date	*18 May*
Work package / Activity *Edge beam to floor slab*			Activity code	*E 1 1 3*
Location	*Workshop / Canteen*		Location code	*2*
Resource	*Materials*		Resource code	3
Type	Description	Quantity	Rate	Total
Concrete	*Edge beam*	*15m³*	*30.00/m³*	*450.00*
Dpm	*Under beam & laps to floor*	*60m*	*0.16/m*	*9.60*
		Total cost £		*459.60*

Figure 5.3 Work package report for materials

Project	*Industrial Development*		Project code	*4 2 3 9*
Week No.	*18*		Date	*18 May*
Work package / Activity *Edge beam to floor slab*			Activity code	*E 1 1 3*
Location	*Workshop / Canteen*		Location code	*2*
Resource	*Subcontractors*		Resource code	4
Type	Description	Quantity	Rate	Total
Subcontract (a)	*Formwork to toe of edge beam*	*30m*	*5 .00/m*	*150.00*
Subcontract (b)	*Mild steel reinforcing bars*	*30m*	*13.00/m*	*390.00*
		Total cost £		*540.00*

Figure 5.4 Work package report for subcontractors

Project *Industrial Development*	Project code *4 2 3 9*
Week No. *18*	Date *18 May*
Work package / Activity *Edge beam to floor slab*	Activity code *E 1 1 3*
Location *Workshop / Canteen*	Location code *2*

Week No.	Planned cost (£)	Actual cost (£)	Variance (£)	Planned progress (%)	Actual progress (%)	Variance (%)
18	*1150.00*	*129.50* *102.00* *459.60* *540.00* ——— *1231.10*	*+81.10*	*20*	*29*	*+9*

Reason for variance / Action taken
Concrete spillage, additional power floating, poor access. Fast formwork contractor.

Figure 5.5 Variance report for work package

To overcome this problem fully is difficult due to the historical nature of invoicing. However, considerable improvement is made by the introduction of the IRS system. Each work package has its own budget including materials and also has its own reporting sheet. Therefore full information on materials for each area can be applied, providing a good guide to the material costs in relation to budgets. In this way the future costs of materials can also be judged leading to a correct statement of profit and clear understanding of the contract status. The IRS reporting system would be weekly. This provides site management with a more up to date view of activity and resource needs at a time before problems occur and the project falls behind schedule.

The decisions that site management make will depend on two major factors, namely previous experience and the information provided by their control system. The IRS system provides a wide range of co-ordinated information on both costs and progress relevant to each clearly defined area for the contract, or work package. This should lead to more accurate and relevant decisions. Variance from the control plan can be accurately assessed and located at the root cause for correction. The availability of historical data adds to the previous experience of the management. The ability to forecast specific events, from data specific to those events, allows genuine control of cost and time, not merely historical confirmation of events.

The case study has shown that there are many common elements between a cost system and a time system: the programme develops dates which are required for calculating cash flow forecasts; the budget estimate provides data for resource outputs required for determining activity durations and resource levels; the cost reporting generates information on work completed and payments made which relate to schedule progress; the remaining activities can be extrapolated to give cost at completion forecasts. The primary aims of integration are to transmit information between the systems, provide better utilisation of data and provide consistency of information. There is a natural dependency between when the work is scheduled and the spending of the budget and the budget can be phased over the time of the project. The effect of contract changes can be more easily found when both the budget and the schedule are involved, as well as improving the accuracy of estimates on completion.

<u>CHAPTER 6</u>

Resource Supply and Control

Site Cost Control Systems

Short term planning and an anticipative, rather than a reactive, approach are essential for site based cost control to be effective. Construction is characterised by one-off, non-cyclic tasks that are carried out over a short period of time, with the result that by the time problem areas are identified it may be too late to correct the activity. Site cost control is essentially for the benefit of site management. Feedback information, though a basic part of the total process, is no substitute for forward planning and organisation. However, to have it in addition to site control enhances the benefit derived from the cost of the system.

Control Systems in Use

Cost control systems used vary from company to company and sometimes from project to project. A survey of construction companies in the UK would probably find that all the companies, regardless of size, are using monitoring systems more than control systems to manage their budget costs. This makes the control system an accounting system whereby budgeted costs are compared after the activities are completed. Most of the companies would be using systems which gave an indication of whether a profit or loss would be made on the overall project. This appears to be the simplest and cheapest system to use. Most companies appear to have a problem in collecting data for their system as this requires money, and sometimes large numbers of staff need to be employed. Preparing reports on a monthly basis, rather than weekly, only helps to reduce the time required for collecting data.

The use of all-in rates to convert hours to money can be used as a basis for control. All-in rates for labour and plant are the estimated total cost per hour as derived by the estimator and included when pricing the project. The advantages of using all-in rates are that cost estimates can be prepared as soon as the work is completed for the week and discrepancies are only due to productivity loss, usually in the control of the site manager. Losses due to fluctuations in cost do not show as a discrepancy as these are beyond the control of the site manager, and there is less data manipulation from pay-roll and invoices to prepare cost statements. The disadvantages are that it may be difficult to reconcile all-in rates used with pay-roll and on-costs are not included and will have to be monitored separately.

A site cost control system is recommended for direct labour (including labour only subcontractors), plant and site overheads. The stages of data collection and analysis are as follows:

- Daily time allocation sheets of operative and plant hours.
- Time and wages weekly cost summary to check allocation and relate to total wages cost.
- Abstract sheets which summarise time against a measure of work (bonus calculations can also be included).
- A feedback record sheet collects similar work together for comparison with a standard rate for labour and plant.

- A cost control statement compares value with cost for a gain or loss for each item or activity.
- Finally a cost control summary sheet reduces the value cost comparison to a trade or work section basis.

Reporting

Reporting is vital to any cost control system as a means of bringing information to each level of management. Reports should contain the results of monitoring performance and their comparison with objectives. Information for control purposes should be compiled on a basis of relevance, accountability, selectivity, yardsticks for comparison, accuracy, timeliness, simplicity and follow-up action.

The report must assist in interpreting the level of achievement and no manager should be presented with information relating to operations over which they have no control. Every manager should know what is actually being achieved and have power to regulate what is happening. The system can be detailed if required to identify precisely any area losing money. Control may collapse if management attempts to control too much, with the really important issues becoming submerged in a mass of irrelevancies. A reported figure in isolation will have little meaning and comparison is necessary for appraising performance. The different levels of management will also require information at varying levels of detail. Facilities should also be available to break down figures into labour, plant, materials, subcontractors and possibly preliminary items. If inaccurate information is presented for decision making purposes, incorrect decisions may be the result.

Reports must be up-to-date in order that any necessary control action can be taken before the events that form the subject of the reports become only of historical interest. The report needs to be easily understood with good communication and coordination between those using the information. In order that follow-up action can take place, items that are going wrong have to be revealed to highlight what has already gone wrong, to determine how and why failure has occurred and to find out who is responsible.

Control Objectives

An operational control system must involve the tasks of setting objectives, reporting, evaluating and taking corrective action. The objectives of such a system are:

- To focus management attention on cost trouble spots in time for corrective or cost-minimising action to be taken.
- To keep each project supervisor informed of the budget for their area of responsibility and how expenditure performance compares to the budget.
- To create a cost-conscious atmosphere so that all persons working on a project will be cost-conscious and aware of how their activities impact on the project cost.
- To minimise project costs by looking at all activities from a cost-reduction point of view.

A control system must relate planning and costing on a short-term or weekly basis. Evaluation of performance should be easily applied by site management which may mean using a simplistic, overall level of detail such as the work breakdown structure. Capturing data should be no more involved than obtaining operative and plant hours worked and comparing with a company or estimating standard for each activity. It must be recognised that prevailing site conditions may require the standard to be adjusted by the site manager in relation to the construction method. Dissemination of information is by the site manager to relevant section foremen or subcontractors. Reconciliation of cost and value for the project as a whole is to remain with the contracts surveyor and will therefore take place on a monthly basis.

When developing a system the use of a fine level of detail in cost control can have undesirable aspects that are to be avoided. Such aspects include too much time being spent on data processing with little resulting management action, too tight a control giving high overhead costs to maintain the system, a fluctuating work performance prompting unnecessary corrective action and a finely detailed work breakdown structure increasing the probability of cost miscalculation.

Subcontractor (or Specialist Contractor) Control

Subcontractors carry out most of the work on construction sites and are responsible for supplying the majority of labour and material resources; they also contribute substantially to the design of their specialisms. Individual firms are part of a diverse and fragmented group of specialists, who in turn work with their own specialist subcontractors, for example carrying out site erection. Most subcontracts will be let on the basis of a concept design, a services line drawing for example. The detailed design, scheduling and ordering of the subcontractor's work usually takes place during the construction phase, unless the subcontractor is part of the design team. Only on appointment will the subcontractor check the initial estimate, with most work requiring design which has to be checked by others and often details have to be changed to fit in with other subcontractors. There is therefore considerable pressure on subcontractors to produce drawings quickly and secure supplies during the construction phase of a project. The variances arising from design and estimating, including unforseen costs, over- or under-spending, will owe nothing to the contract and the efficiency with which it is executed. Financial control is based on the contract sum and cost control on the actual work as carried out on site.

It is usually the responsibility of the subcontractor to contact the site to agree start dates and a programme of work. Delays in the readiness of sites tend to increase with the completeness of the building with many subcontractors expecting a delay as a matter of course and are often caught out if work on site goes according to plan. Supervision of subcontract operatives depends on the size of the project, with smaller less skilled trades left to the supervision of the main contractor. Most labour is paid on piecework rates and the ability of the main contractor to provide continuous work on site is important.

Relationships on site depend on the level of competence of site management and the ability of the subcontractor to complete work of the required standard on time. Relationships become strained if the main contractor is not seen to be in control of the building process and lacks confidence in preparing a firm programme with adequate notice. Subcontractors are increasingly being asked to take responsibility for complete elements of the building and do so by employing specialists to work directly for them. Subcontractors are also expected to be more self-sufficient on site, taking responsibility for plant, access, setting out, and the like.

Control of subcontractors includes planning and scheduling work, identifying key procurement items, distributing the value of work, measuring physical progress, determining value earned and taking corrective action in relation to time, cost and quality performance. The required attributes of a subcontractor cost control system which can be operated by site management are as follows:

- The minimum of data should be used to overcome resistance to collection and data should be optimised by standardisation in breakdown and reporting.
- Any system should reflect the obligations of the main contractor and the subcontractor with respect to planning and control.
- The system should enhance communication between the parties concerned.
- A minium number of forms should be used with variations and claims consistent with the initial breakdown.
- Any analytical technique used should be familiar to the users, that is the main contractors and the subcontractor's site management.

- The system should be usable on a day to day basis to reflect required performance monitoring.
- A forward looking, problem solving emphasis should be available to identify future options and documentation should be available to assist in the resolving of claims.
- The system must be easy to update to encourage its use.
- Both overall control and detailed control information should be presented simultaneously.
- Progress in terms of quantities placed and money expended should be presented by visual means.
- Feedback of unit costs and production rates in relation to a work breakdown structure should be available for future use.

The accumulation of a database from such a control system, structured in accordance with information required for estimating and planning, will enable project information to be shared with all company personnel. The system will also meet the control objectives previously established.

Operation Execution Systems

If a system is to be successful a company will have to ensure that the system is acceptable at site level, including its application to subcontractors. There is a problem in that senior management has a different set of requirements to site based personnel. When considering the objectives of construction project planning systems it is often found that control is the overriding factor. Most documents produced by planning systems illustrate control in both their format and content. Project plans are not viewed by senior management as 'living documents' and management functions are usually preoccupied with obtaining forecasts and control reports. The planning departments in many companies are termed 'control groups'. One result of this process is that while the formal plan (the project network, for example) is used by the project manager, the actual execution of the works is based on informal short-term plans prepared by site staff. This approach may give rise to problems if little or no regard has been paid to the formal plan. Resources with long lead times are not available when required and optimising planning is not possible. Integrating the various project components becomes difficult and maintaining consistency between multi-level decisions is not feasible.

If there is an emphasis on control rather than execution, this can lead to 'retrospective' planning, which is directed towards removing deficiencies produced by past decisions. A greater emphasis on execution rather than control, as in 'prospective' planning, will lead towards creating a desired future. It has been suggested that top management's inability to produce changes at site level may cause a reversal of the control process. Rather than taking steps to correct performance variances, most variances are remedied by adjusting the plan to actual performance. Computer aided MIS may make this problem more acute where large complex networks are produced. The plan can become absolute and as a result inflexible and bear less resemblance to reality. Work packages are planned to a programme. Their commencement and completion are stages of progress and the resources applied, as well as the duration, are checked at interim periods if they extend over more than one accounting period. If a network has critical work packages then there must be a reason and the network requires updating. Updating will utilise historical data which reinforces the control role of the process. Therefore any computer aided MIS should utilise a feed forward as well as a feedback system.

Case Study

A detailed system for subcontractor control can include a subcontractor price breakdown, a subcontractor schedule and progress chart, and a procurement status report which are all used at the commencement of the subcontract to establish basic data. During the period of the subcontract a weekly subcontract report and a subcontract progress claim are used. Forms based on the system are shown in Figures 6.1 to 6.5. The subcontractor price breakdown identifies relevant activities and stages, assigns prices, computes subtotals and total cost, and identifies quantity descriptors, with all data being agreed between the parties concerned. The subcontractor schedule and progress chart determines a scheduled start and finish date, facilitates short term planning, measures performance, and updates the programme. The procurement status report progresses the requirements for selected work materials. The weekly subcontractor report assists with labour requirements and productivity information. Finally the subcontract progress claim computes the percentage complete for each activity, the value to date for each activity, the total value to date, and current payment due.

Some general benefits from using such a system are that the detail of information provided lessens the adversarial nature of relationships, particularly when the subcontractor sees that the system could be of benefit to themselves. The system could also be used for a wide variety of subtrades. Important associated issues are seen to be standard activity coding systems, the interface between the main contractor's master plan and subcontractor's schedules, and any contractual implications in the use of the system. Once cost has been assigned to work packages, an integrated computerised management system can automatically calculate payments due to subcontractors once progress has been entered. This is an advantage of having an activity based planning system as the integrative mechanism and shows that it is feasible to output data from a construction site that monitors and integrates the functions of both planning and job cost accounting.

Subcontractor Price Breakdown

Project .... Date ....
Subcontractor...
Specialism .... Initial breakdown
Contract No. .... or Variation No. ....

	Activity breakdown				
	Formwork		Concrete		
Location	£	Qty.	£	Qty.	£
Basement	4500	3500m^2	2450	70m^3	6950
Grd. floor	5700	4000m^2	3400	85m^3	9100
1st. floor	3100	3400m^2	2630	75m^3	5730
2nd. floor	2250	3400m^2	2630	75m^3	4880
Overheads	1300		950		2250
Totals	16850	14300m^2	12060	305m^3	28910

Approved by .... Date .... (for main contractor)
Approved by .... Date .... (for subcontractor)

Figure 6.1 Subcontract price breakdown proforma with activity breakdown as per standard work breakdown structure.

Subcontractor Schedule and Progress Chart

Project .... Progress date ....
Subcontractor ....
Specialism .... Key: Plan start date Plan finish date
Contract No. .... Actual start date Actual finish date
Progress percentage

Location	Activity breakdown: Formwork		Concrete		Notes
Basement	040698 120698 xxxxxxxxxx100%	250698 030798	110698 190698 xxxxxxxxxx100%	250698 030798	Completed as shown in key above.
Grd. floor	290698 030798 xxxxxxxx80%	310798	060798 100798 xxxxxx60%	310798	
1st. floor	030898 170798 x10%	210898	030898	210898	
2nd floor	240898	110998	240898	110998	

Figure 6.2 Subcontractor schedule and progress chart with activity breakdown as per subcontract price breakdown with planned dates taken from the master programme.

Procurement Status Report

Project .... Date ....

Key: **P** = Planned date
D = Duration
A = Actual date

Item	Quantity	Key	Prepare documents	Tender call	Contract award	Submit drawings
Curtain wall	200m^2	**P**	060798	200798	170898	240898
		D	2 weeks	4 weeks	1 week	3 weeks
		A	100798	240798	280898	

	Approve drawings	Fabricate	Delivery	Inspect	Variance
P	140998	210998	021198	161198	
D	1 week	6 weeks	2 weeks	0.5 week	2 weeks
A					behind

Figure 6.3 Procurement status report with variance indicating if ahead or behind procurement schedule.

Weekly Subcontract Report

Project .... Report date ....
Subcontractor ....
Specialism ....
Contract No. ....

Activity	**Number of operatives**								Location and progress
	M	T	W	T	F	S	S	Total	
Formwork	8	8	8	10	10	6	-	50	Grd. floor 80% complete
Formwork	4	4	4	2	2	2	-	18	1st. floor 10% complete
Total operatives	12	12	12	12	12	8			
Total supervisors	2	2	2	2	2	1			

Approved by .... Date .... (for main contractor)
Approved by .... Date .... (for subcontractor)

Figure 6.4 Weekly subcontract report with activity breakdown as per subcontractor price breakdown and showing level of supervision.

Subcontractor Progress Claim

Subcontractor ....
Specialism .... Progress claim No. ....
Contract No. .... Period ending...

	Activity breakdown Formwork			Concrete			
Location	Total value	% compl -ete	Value to date	Total value	% compl -ete	Value to date	Total value to date
Basement	4500	100	4500	2450	100	2450	6950
Grd. floor	5700	80	4560	3400	60	2040	6600
1st. floor	3100	10	310	2630	0	0	310
2nd. floor	2250	0	0	2360	0	0	0
Totals	15550		9370	11110		4490	13860[A]

General overheads	400[B]
Value of variations to date	550[C]
Value to date [A+B+C]	14810
Less retention	740
Less previous claim	6430
Current payment due	**7640**

Approved by .... Date .... (for main contractor)
Approved by .... Date .... (for subcontractor)

Figure 6.5 Subcontractor progress claim.

CHAPTER 7

Coordinated Project Information

Introduction

There is an incompatibility between estimating and the production process together with any relevance it has to resource control. In competitive tendering the completed building should reflect the information provided at tender stage but rarely does so. Any calculation of resources at the start of a contract will therefore remain fluid for much of the duration of the building process. Items like provisional quantities, provisional sums, contingencies and possible variation create gaps in the information available to the contractor. Resource control is seen to involve labour, plant and materials, but information is also a resource. It is in the form of drawings, specifications and instructions and requires control. A limited proportion of resources are in the direct control of the site manager when there is a high element of subcontracting.

The direct cost of production (labour, plant and materials) plus the indirect costs (site staff, accommodation, and the like) can be readily derived, but the detail is in different forms from that in which resources are developed or consumed. It is impossible to provide any degree of control without the provision of good budgetary information and the problem is one of quantifying needs. Those reflected in the documentation of the contract are often in the wrong terms.

Budgetary control systems that give good information and assist and influence senior management in decision making can surely be devised. Control requires the definition of objectives. Resource requirements have to be established before operations start. Adequate drawings and information are prerequisites. The flow of subcontractors' drawings and information will also have to be monitored, particularly their checking and approval. The principal skill in production control will lie in procedures that enable site managers to apply their resources flexibly in a changing environment while amendments to drawings, their dates, details and implications are of particular concern.

Concept Models of Buildings

Progress towards a coordinated information system will depend on the development of a coded database that links data within the system. An overall aim in establishing concept models for building is to relate a basic framework for information processing systems for the whole building process. Criteria for the future use of concept models to meet this aim are as follows:

- A faceted classification system which defines the total building concept in order to improve data exchange between users.
- An adaptable model that relates to both the stages of a project and to the differing needs of the users.

- A system that realises the interrelationship between spatial dimensions, quality and quantity statements.
- The provision of a project database that facilitates individual specialist use at any level.
- A coded cross-referencing system to coordinate computer programs.
- The integration of alpha-numeric and graphical data in a Database Management System that can be used top-down or bottom-up.
- The facility to use activity descriptions as a basis for production control.

Coordinated Information Systems

The requirements for project information are now considered in relation to work carried out on the coordination of information. The structure of such a holistic system should include all design and build activities. The method used to procure work by the client will also influence the way in which information is produced and how resources and services are bought in for a project.

Information Systems

Construction projects are systems of coordinated teamwork. Information systems consist of words, numbers and symbols which relate to physical objects and processes. Initial project objectives are derived from the client's brief. They are interpreted in a model comprising drawings and specifications. A description or model of the project organisation is also developed in parallel. This expresses production objectives in terms of physical and financial plans.

An information system distributes information about a project to those involved in its organisation. It should do this in a convenient form and at appropriate times. The complexity of project information when each team produces data to suit its own requirements independently of others can be greater than it need be. Information systems should bring order to construction project information. An ideal system should enable the teams that make up the project organisation to record all their decisions and progress, while being able to select and examine what they require from the total set of project information.

Computing facilitates the integration of design and management information, from which feedback on the behaviour of the project could then enhance the precision of planning and estimating. A computer based information system should also increasingly cope with routine matters, leaving managers free to concentrate on decision making. No amount of coding and computer processing can completely compensate for a flawed structure of the information system.

The increasingly more complex construction process has been recognised. There are sub-consultants in the design team as well as subcontractors in the building team, all of whom have to contribute to the end result at different times. The constraints of timing and the financing of a construction project all have their influence on the need for flow of information between the parties involved. Documents should be produced which satisfy the needs of the contractor, facilitate retrieval of information by each member of the building team and encourage the production of a well integrated design by providing consistent documentation.

Data Coordination

Data coordination should lead towards a reduction in time for the design and build process and the possibility of completion within time and cost targets. Feedback of experience and tried design details should be better, resulting in trouble free construction and buildings. The more versatile documentation should give the contractor a better chance to organise precontract planning and contract administration more effectively. These claims are considerable and no improvement will be gained without effort from those introducing data coordination.

When the information provided to contractors is insufficient, conflicting or incorrect, problems occur on site which lead to a consequent reduction in the quality of work, delays and increased costs. Coordinated Project Information (CPI) has provided a Common Arrangement of Work Sections (CAWS), the purpose being to define an efficient and generally acceptable building works arrangement for specifications and bills of quantities for building projects.

CAWS is replacing existing methods of arranging specifications and quantities in the UK. It has been used as the basis for The Standard Method of Measurement of Building Works, 7th edition (SMM7) and the National Building Specification has been restructured to match. The main advantages should be:

- Easier distribution of information to subcontractors and suppliers.
- More effective reading together of documents where work is divided into smaller sections giving reference between measured items and the specification.
- Greater consistency of content and description by using standard specification clauses with the amount of information given in project documents being less variable.

About 300 work sections are included that have been derived from observation of current practice and follow the pattern of subcontracting. The main factors influencing work groupings are responsibility for design and performance and methods of working in relation to subcontract practice. The fundamental concept of the work section is that of a division of building work for purposes of procurement. Specifications and bills of quantities are taken to reflect construction activity which then requires the addition of skills, knowledge and the use of specialist equipment to the resources of labour and materials.

The arrangement of work sections into recognisable and consistent groupings is achieved by using three levels as follows:

Level 1	Group	eg. R	Disposal systems
Level 2	Subgroup	eg. R1	Drainage
Level 3	Work section	eg. R12	Drainage below ground

Levels 1 and 2 are headings under which the work sections are grouped. An overriding consideration was simplicity with short and easy to remember section numbers. The logic of the sequence of headings is given in the Appendix in the Work Section Master List.

Programme activities can be restated in terms of CAWS and reflect the division of work between different subcontractors and work gangs. Cross-reference to the CAWS section makes each activity traceable back to the relevant specification requirements, quantities and monetary values. The programme and any time and cost control system should become an extension of the coordinated project information. The CPI coding is effective at defining work activities by trade or by material, but does not include a reference to location.

Information Integration

A complete assessment of performance is required that interrelates specification, in terms of the materials and technology applied, with the main resource of labour, plant and administration used by the industry. Monitoring the site process can be the first stage with collected data integrated to provide feedback to both design and construction. Such an integration requires an 'open' classification system together with the development of appropriate computing facilities. The classification system must contain the data required by the industry in a flexible mode to serve varying requirements. CPI is criticised as a 'closed' system which may only have short-term benefits.

The classification system that is recommended as the basis for integration is based on the Construction Indexing Manual (CI/SfB) originally developed for classifying project drawings. The system is faceted in design, which suits computer supported data files and their free sorting, and the system was made 'open' so that, for example, project information for any specific operation can be assembled as one document. An example coding is (21) representing Superstructure, External Walls and can be seen in the Appendix in the Elements Master List. The successful integration of project information will depend on the data developed in design being contained in a classification system that can be continued into contracting to contain the project information required for control.

Work is ongoing in developing information systems, but tends to be fragmented by professional or commercial interests. Any coding proposals should therefore take into account existing developments across the total design and construct process in anticipation of future coordination. Conclusions drawn from previous work carried out in the area are that:

- A framework of classifications exist that enables project information to be presented in a format appropriate to any user of project information.
- Documents are available that offer information in an appropriate form for the primary purposes for which they have been designed.
- Individual tasks within the process of tendering have different requirements for project information in terms of the classification of data and the content and level of detail that has to be quantified.

The wider issues of coordinating information systems have been considered for both design and contractors' work, with implications for integration and retrieval. It can be seen that the basis for a national system has been developed and requires consideration when looking in detail at data coding for operational control. It is possible, in terms of database management systems, to present information in an appropriate format to meet the requirements of all users.

Activity Based Schedules for Control

It is recognised that, given forms of contract where contractors are required to produce their own data, it is possible to apply aspects of operational control unconfined by bills of quantity. The aim is to improve the effectiveness of computer based control and information transfer in contracts where the contractor originates basic data on measurement, time and cost. The use of activity schedules as a control and information source is then worthy of development. The aim would be to analyse if they create a greater awareness of progress; ease the task of interim measurement and valuation; assist in assessing the impact of variations on time and cost; and facilitate in the valuation of variations.

Production control requires the comparison of the direct and indirect costs of production with the cost limits or values set by the business unit. This will require budgetary information to be quantified in terms of production activities. Activity based programmes of work also create an awareness of progress in relation to time. They might also ease the task of valuations and allow for the variations of work content, only if the terms in which valuations and variations are priced are activity based.

Analysis of Coding Systems

An information system has to combine the description of the subject with a model of the variables relating to it. In terms of project organisation these would be expressed as time, cost and construction method. A framework of classification should be used that enables project information to be presented in a format appropriate to its users. Those who estimate the costs of the work or plan its progress require data according to the description of work on the one hand and the activity to be undertaken on the other. If these are different then different outputs are required. However, for any subsequent comparisons, they should aggregate at some stage to a single item of data. This aggregation has to be at a useful level of detail. For construction, a coding system can then relate to and incorporate existing systems, for example, CI/SfB for drawings and SMM work sections for quantities. These already formulate the classification of design information and the basis of pricing respectively.

Construction operations can be defined as activities, the time and resource consuming parts of a project, used for the purpose of time and cost control. Activities may be further grouped to classify by finished product, by material or by function. A lexicon and syntax structure are also required to describe activities that are generally recognised and acceptable. All resources might then be encompassed, including both direct and indirect costs, that are to be associated with any project. Data coding and phraseology are therefore available to develop a coding system in terms of an activity based planning system.

Criteria for Coding Systems

Activity codes should be simple, concise and easy to interpret, with only essential and readily identifiable activities used. The main criteria for the analysis of coding systems that are met by the Master Lists are summarised as follows:

- Provide feedback on production performances.
- Classify all essential production activities so that variation in performance can be correlated with their location and situation.
- Classify resources applied in performance (labour, materials, plant, subcontractors and overheads).

- Be consistent in application from activity to activity, job to job and contract to contract, over a period of time.
- Give database capability, through which to research values and particularly variations in value.
- Minimise coding discrimination at point of input.
- Be stratified in structure for reporting at various levels.
- Be faceted in structure for transfer and comparison of information.
- Be related to existing, nationally recognised coding systems.
- Provide a computerised data file.

The coding system represented by the Master Lists, included in the Appendix, supports the view that output data from a site can coordinate control data. Classification will involve the use of a coding system that relates design and build information. It will be advantageous if the coding system developed for operational control relates to the types of models used for a coordinated information system.

Case Study

Coding must provide uniformity for transfer and comparison of information between projects while including all items of building construction, although only a few items may be used on any one project. The code must also be capable of being applied in a consistent way from project to project. The structure of the coding system requires to be stratified for overall (e.g subcontractor) or detailed (e.g. material) control, with reporting taking place at weekly or monthly intervals. A molecular coding system that is capable of synthesis will reduce the number of items that have to be allocated codes on site. The coding hierarchy being a faceted structure should suit computer supported data files and their sorting.

The coding order of element followed by work section is derived from the elemental dominance of other coding systems. To prevent the reordering of programme activity descriptions a hierarchy is devised in Figure 7.1. This will follow the thought process normally used in building up an activity description. For example, 'blockwork to internal office' is coded as G3 (blockwork), 22 (internal walls) and 2 (office).

1. Division of company		Printed heading on control	
2. Project		records	
3. Work section	)		
	)		Taken from a
4. Element	)	Cost Centre	schedule for
	)		comparison
5. Location	)		between jobs.
6. Resource	(	Labour	
	(	Plant	Allocated
	(	Materials	at site
	(	Subcontractors	
	(	Site overheads	
	(	(preliminaries)	
7. Process	Not required with new work		Taken from a schedule.

Figure 7.1 A coding hierarchy shown in order of application.

A Master List of element coding based on the CI/SfB Project Manual and work section coding based on SMM7 has been developed for cost centres. The Master Lists are included in the Appendix. An application showing the classification and coding of programme activities is included in Figure 7.2 using two digit work section and element coding only.

Activity	Code
Demolition	C 1 1 1
Reduce level excavations	D 2 1 1
Piling	D 3 1 7
In situ concrete foundations	E 1 1 6
Drainage	R 1 5 2
Hardcore	D 2 1 3
Steelwork frame	G 1 2 8
Metal sheeting to roof	H 3 2 7
External brickwork	F 1 2 1
In situ concrete ground floor slab	E 1 1 3

Figure 7.2 Coding activity descriptions.

CHAPTER 8

Modelling Operations

Simulation

Simulation is the representation of features of the behaviour of a physical or abstract system by the behaviour of another system. Queueing theory has provided general solutions for simple situations, but there are few such situations in construction to which it can be applied. A better appreciation of the situation may be gained by introducing a dynamic element in Monte Carlo simulation, a method of obtaining an approximate solution to a numeric problem by the use of random numbers. The application of Monte Carlo simulation to construction problems, which involve queues of all types, is where a dynamic model of a situation is produced from observations of that, or a similar, situation in order to predict the likely operation of the subject being modelled over a period of time. Simulation is used to carry out rapid cycles of experimental work in situations where durations are long and/or patterns of behaviour vary from cycle to cycle within bounds. The analysis of historic data can be translated into predictions of future behaviour, permitting the comparison of alternative courses of action in sufficient time to alter strategic and not just tactical decisions.

The general approach is to simulate a real life situation from random samples of activity, relating occurrences or intervals to a probability distribution curve. This approach contrasts with the average times and average rates that are the more customary sources of estimated times and costs. The times required by the processes that make up construction operations are not constant, being dispersed about a mean value. The probability that they will differ from that mean value by some amount can often be estimated. The basic factors that influence work task durations, and therefore construction process durations once the operational technology has been decided, are process design, work content, equipment, labour, working conditions and efficiency of management. The influence of these factors on work tasks varies between construction sites. Therefore, work task durations may be deterministic, probabilistic around a known mean or almost completely random.

The outcome of a situation is deterministic if it is certain, having selected or knowing all the variables involved. A deterministic duration assigns a specific fixed value to the duration of the work task. The duration is deterministic because a specific time is required, a specific mean value has only small variations and a constant duration can be taken over a period of time or the purpose of the model is such that probabilistic variation can be ignored. Construction operations taking place in a controlled environment may therefore use a single value of time duration when the variability is considered small or insignificant. A deterministic analysis of system productivity may also be undertaken because it is sufficiently accurate for the purpose of the researcher, but the loss of productivity due to the random variation of work cycle durations may be significant enough to justify consideration.

In situations where the values of some of the variables are not known, and an allowance has to be made for the varying probability of certain values, the situation is probabilistic or stochastic. When stochastic models are used it is necessary to select values of the variables which are representative of the available data. Then, providing there is some measure of the frequency of occurrence of certain values, random sampling may be used to postulate the likelihood of an event. The use of random sampling ensures that every item in the population has an appropriate chance of being selected. Random numbers can be obtained from statistical tables or can be generated by calculator/computer.

Random durations in the movement or flow of resources can, however, delay work cycles and operations by increasing the time that resource units spend in idle states awaiting release to productive work tasks. Probability distributions are useful in describing observed variation of work task times on site, since they are easily defined in a mathematical sense by relatively few parameters. Probabilistic functions are then used to define the populations from which random time durations can be taken for the simulation of a system. Random variates can then be generated by entering the cumulative probability function. Table 8.1 shows an example of the build-up of cumulative distribution from site observations.

Table 8.1 Site observations of fixing times for precast cladding panels.

Interval minutes	Number of observations	Relative frequency or probability density	Cumulative number of observations	Percentage cumulative distribution
00 - 04	02	0.04	02	04
05 - 09	20	0.36	22	40
10 - 14	25	0.45	47	85
15 - 19	03	0.05	50	90
20 - 24	04	0.06	54	96
25 - 29	02	0.04	56	100
	--	----		
Total	56	1.00		

The object of developing a model of what is known as discrete unit simulation for a production system is to:

- examine interaction between flow units,
- determine idleness of productive resources,
- locate bottlenecks, and
- estimate production of the system as constituted.

Some of the questions by which a simulation can be tested for validity before its conclusions are applied are as follows:

- Does the simulation have a low variance of outputs (internal validity) when repeated with all the external inputs constant?
- Does the model read reasonably when manipulated by the researcher or the real life decision maker?
- Do the relationships between the variables within the simulation correspond to those in real life (a quantification of previous question)?
- To what extent does the simulation enable one to predict observable events or event patterns?
- Do the relationships between observations within the simulation correspond to the same relationships between observations in the real world (hypothesis validity)?

The validity of any simulation is a function or result of how much is known of the statistical distribution of the actual construction operation times. Only adequate sampling or records can establish a true distribution.

The importance of recognising the limitations of simulation is that the attempt to simulate real life situations is limited by the parameters set by the researcher. Simulation can be reliable in comparing alternatives, such as the productivity of two proposed systems, but accuracy in predicting actual output or cost for a particular system may be limited by the accuracy of the input data and the length of simulation run. But simulation is a very flexible and useful tool that, if used appropriately, can generate a great deal of information about the system studied. It is also a technique that, without violating the nature of reality too much, is easiest to understand and to adapt to actual situations.

Simulation Application

Resource planning by construction managers depends on a combination of experience and analysis, particularly if standardisation of design permits work to proceed in a repetitive manner. If a repetitive unit project requires in-depth planning analysis to maximise the productivity of labour and plant, simulation modelling can be a valuable quantitative technique. Project management must be concerned with controlling costs and adhering to a time schedule, as well as the construction process, and maintaining a continuous flow of work may bring benefits from experience learning curve effects.

The recognition of repetitive cyclical processes when modelling construction operations is best developed using a computer language, GPSS for example, to facilitate the simulation of construction operations on large scale operations. Operations, therefore, with a repetitive and cyclical nature make good subjects for modelling as there is a need to ensure an efficient cycle to reduce the effect of small inefficiencies being multiplied many times. It is also suggested that simulation may produce more accurate results than other techniques because it is less dependent on invalid assumptions, but that modelling operations by gathering new data in each case is time consuming and costly. Perhaps the most important benefit is that experimenting with a model of a construction system can give a better understanding of the system in relation to changes in resource allocation, the decisions made and the influence of external conditions. But the reservation has to be made that the control of construction activities today has not yet advanced to a state where a high degree of sophistication in analytical method can have significant benefits, except in highly mechanised processes.

An operations manager has to forecast the results that he would hope to obtain in practice and average historic production feedback is not always an adequate guide to future performance. One of the most important problems in construction work is that of balancing resources, there being two types of resource: (a) cyclical, usually associated with plant, and (b) that with a random or one-off pattern of behaviour as typified by labour processes. Construction processes should be measured and the ranges of productivity recorded rather than average readings. The total time for a work cycle can then be found by random sampling from frequency distributions of each part of the cycle and by running such a model many times practice can be realistically simulated. The basic considerations that are required before simulation becomes practical are that the data is adequate and correct, and that the situation is sufficiently complex to make the exercise worthwhile. Simulation should only be regarded as a technique to be used alongside others, such as network analysis in the planning, progressing and resource allocating problems. Also for better understanding it is initially desirable to carry out simulation by hand, although simulation could best help in a complex situation and should be used for planning resources at the estimating stage rather than as a site planning method.

It would appear from the recommended applications of simulation that it may well be best applied, when suitable to do so, to augment the planning that takes place at the estimating or pre-contract stage for large and complex projects. If the construction manager wishes to use simulation on site it may well be in a limited form, within criteria previously set by the estimating and planning functions of the construction contractor, although if data is gathered on a particular site, the frequency distribution so collected will be of particular significance to the unique situation that exists on that site.

Case Study One

A low-rise office block had an in situ concrete frame with large concrete pours being made into suspended waffle floor slabs. Data collection for the first pour was made using work sampling with the object of predicting the time of future concrete pours. The frequency distribution for the inter-arrival time of ready-mix concrete trucks is shown in Table 8.2 and the service time in placing the concrete pump is shown in Table 8.3. Twenty simulations are then carried out by hand as illustrated in Table 8.4, resulting in an overall time of 8 hours and 7 minutes for a 100 m^3 concrete pour.

Table 8.2 Time distribution for ready-mix concrete trucks.

Inter-arrival times (minutes) for two ready-mix concrete trucks (5 m^3 each)	
Mid point of range	Cumulative percentage frequency
03	06
09	19
15	54
21	67
27	76
33	82
39	91
45	97
51	100

Table 8.3 Time distribution for placing concrete by pump.

Service time (minutes) for concrete pump on trucks.

Mid point of range	Cumulative percentage frequency
03	02
05	32
07	52
09	71
11	80
13	86
15	91
17	94
19	99
21	100

Table 8.4 Monte Carlo simulation for a 100 m^3 pour.

Simulation number	Random no.	Inter-arrival time	Time of arrival	Random no.	Service time	Time, service ends
1	93	45	45	72	11	56
2	81	33	88	02	03	91
3	79	33	121	66	09	130
4	25	15	136	28	05	141
5	64	21	157	58	09	166
6	96	45	202	96	19	221
7	66	21	223	22	05	228
8	02	03	226	32	05	231
9	30	15	241	28	05	246
10	59	21	262	73	11	273
11	30	15	277	26	05	282
12	24	15	292	80	11	303
13	32	15	307	97	19	326
14	14	09	316	08	05	331
15	53	15	331	32	05	336
16	97	45	376	85	13	389
17	76	27	403	70	09	412
18	64	21	424	72	11	435
19	98	51	475	03	05	480
20	03	03	478	39	07	487

Total time is 487 minutes or 8 hours and 7 minutes.

Detailed Modelling

A construction operation can be characterised by its technology, resource use, breakdown into work tasks and sequences and the nature of its ultimate assembly and use. The best technology available has to be selected and the relevant construction method defined for each operation. The nature and working of the construction operation must be described in such a way that foremen and operatives may perform the site operation properly and efficiently. Finally progress and control have to be organised in relation to the changing conditions during the performance of the site operation.

In establishing a system for analysis of construction operations a clear and readily understandable terminology must be defined. A work task is the basic descriptive unit in construction practice and is a readily identifiable component of a construction process or operation. A work assignment is a collection of work tasks assigned to an operative for performance and usually involving sequences of work tasks and therefore may constitute a construction process. A construction process is the placing of a definable piece of construction involving some technological process and work assignment structure. It is an expression of the construction method and linked to part of the completed project, a synthesis of construction processes.

Construction operations may be grouped into different classifications, depending on which features are selected: (a) by finished product, (b) by material, or (c) by function. The full description of construction operations requires the identification of the construction technologies involved, the sequencing of the various work tasks that make up the processes and the allocation of the required resources for the work tasks and processes of the operation.

Case Study Two

A model can be developed using the basic modelling elements suggested by Halpin and Riggs, Design of Construction and Process Operations, Wiley, 1992 and defined in Figure 8.1. The amount of detail with which to describe the technology and operations used will depend on the use of the model. An outline model for the fixing and striking of formwork to a waffle floor slab is given in Figure 8.2. The breakdown of operations into work tasks can produce an easily understandable model if detailed control of formwork fixing is required, but in this particular case study there were many different approaches within the overall operational sequence with various sub-gangs carrying out tasks. It may be conjectured that detailed organisation could have led to more efficient working, rather than leaving the carrying out of work tasks to the operatives' discretion. Data can also be collected for the frequency distributions of the work tasks and the construction process simulated as shown in Case Study One.

Modelling Element	Name of Element	
	Normal	The normal work task which is unconstrained in its start and indicates active processing of (or by) resources.
	Combi	The constrained work task which is constrained by a combination of resources in its start, otherwise similar to the normal work task.
	Queue	The idle state of resources, representing a queueing or waiting of resources.
	Arrow	The directional flow of resources.

Figure 8.1 Basic modelling elements for CYCLONE.

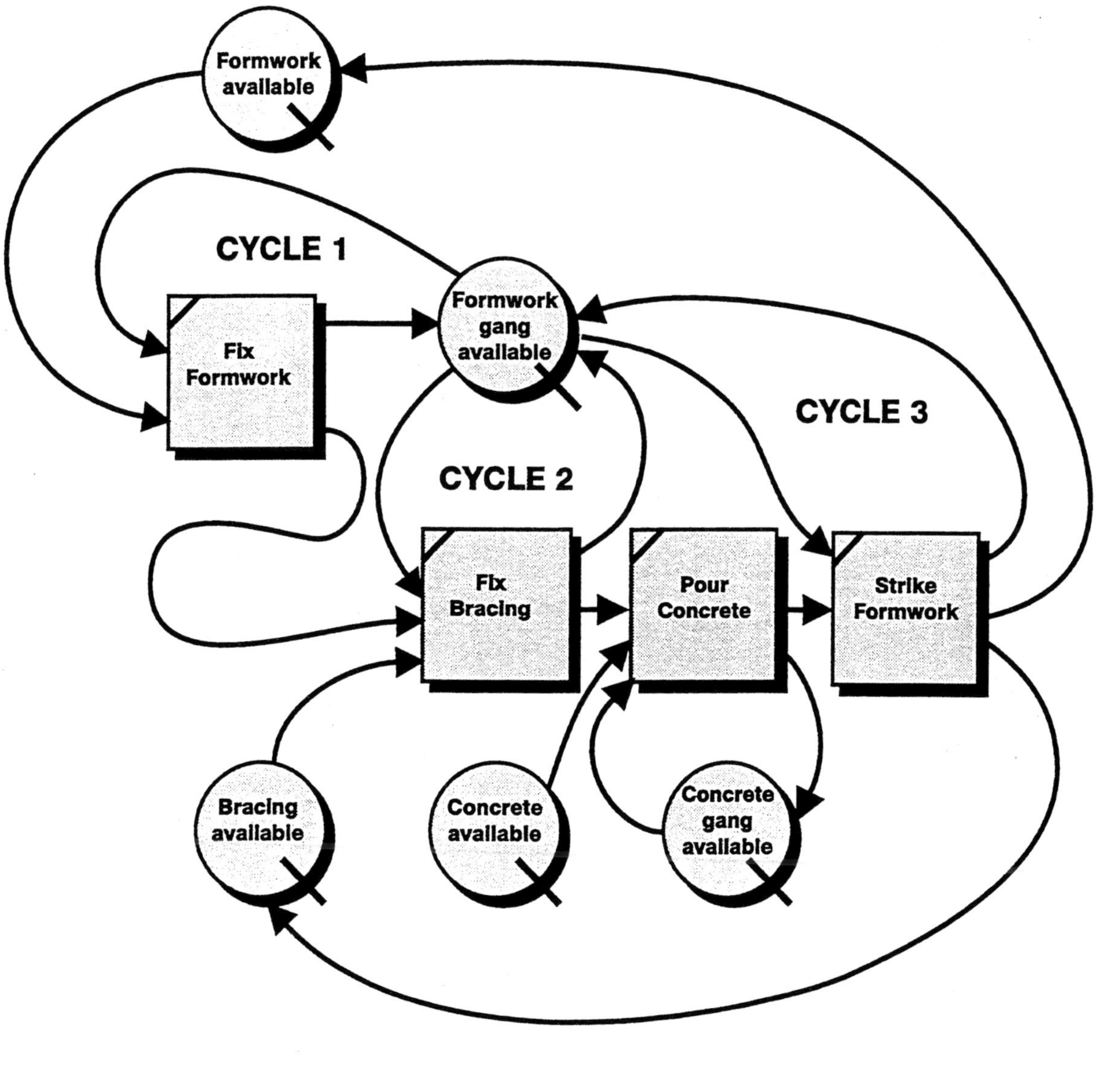

Figure 8.2 Model of formwork operations to waffle floor slab.

APPENDIX

Master Lists for Classification of Programme Activities

Elements

Classified as CI/SfB Project Manual (RIBA) Table 2 by two digits

1	-	SUBSTRUCTURE
1	0	Site substructure
1	1	Ground excavation, earth shapes
1	2	Free
1	3	Floor beds
1	4	Free
1	5	Free
1	6	Foundations
1	7	Pile foundations
1	8	Reserved for other substructure elements
1	9	Summary

2	-	SUPERSTRUCTURE
2	0	Site superstructure, primary elements
2	1	External walls
2	2	Internal walls, partitions
2	3	Floors, galleries
2	4	Stairs, ramps
2	5	Free
2	6	Free

| 2 | 7 | Roofs

| 2 | 8 | Frame

| 2 | 9 | Summary

|| 3 || - || SUPERSTRUCTURE, SECONDARY ELEMENTS

| 3 | 0 | Site superstructure, secondary elements

| 3 | 1 | Windows, external doors (secondary elements in external walls)

| 3 | 2 | Internal doors, hatches (secondary elements in internal walls)

| 3 | 3 | Access floors, traps, stages (secondary elements in floors)

| 3 | 4 | Balustrades (secondary elements in stairs, ramps)

| 3 | 5 | Suspended ceilings (secondary elements under floors and roofs)

| 3 | 6 | Free

| 3 | 7 | Roof lights, etc. (secondary elements in roofs)

| 3 | 8 | Reserved for other secondary elements

| 3 | 9 | Summary

|| 4 || - || FINISHES

| 4 | 0 | Site finishes

| 4 | 1 | External wall finishes

| 4 | 2 | Internal wall finishes

| 4 | 3 | Floor finishes

| 4 | 4 | Stair ramp finishes

| 4 | 5 | Ceiling finishes

| 4 | 6 | Free

4	7	Roof finishes
4	8	Reserved for other finishes
4	9	Summary
5	-	SERVICES (PIPED AND DUCTED)
5	0	Site services
5	1	Refuse disposal
5	2	Drainage
5	3	Hot and cold water
5	4	Gas, compressed air, etc
5	5	Refrigeration
5	6	Space heating
5	7	Air conditioning, ventilation
5	8	Reserved for other services
5	9	Summary
6	-	SERVICES (ELECTRICAL)
6	0	Site installations
6	1	Electrical centre
6	2	Power
6	3	Lighting
6	4	Communications
6	5	Free
6	6	Transport

6	7	Free
6	8	Security
6	9	Summary
7	-	FIXTURES
7	0	Site fixtures
7	1	Circulation fixtures
7	2	General room fixtures
7	3	Culinary fixtures
7	4	Sanitary fixtures
7	5	Cleaning fixtures
7	6	Storage fixtures
7	7	Free
7	8	Reserved for other fixtures
7	9	Summary
8	-	LOOSE EQUIPMENT
8	0	Site loose equipment
8	1	Circulation loose equipment
8	2	General room loose equipment, furniture
8	3	Culinary loose equipment
8	4	Sanitary loose equipment
8	5	Cleaning loose equipment
8	6	Storage loose equipment

8	7	Special activity loose equipment
8	8	Reserved for other loose equipment, furniture
8	9	Summary
9	-	EXTERNAL WORKS
		External works may be shown separately, if not grouped under 10, 20, 30, 40, 50, 60, 70 or 80

Work Section or Trade or Construction Form or Material

Classified as the Common Arrangement of Work Sections for building work (CAWS) and SMM7

A:	Preliminaries/ General Conditions	A	1	The project generally
		A	2	The contract
		A	3	Employer's requirements
		A	4	Contractor's general cost items
		A	5	Work by others or subject to instruction
B:	Complete Buildings	B	1	Proprietary buildings
C:	Demolition/ Alteration Renovation	C	1	Demolition
		C	2	Alteration - composite items
		C	3	Alteration - support
		C	4	Repairing/renovating concrete/masonry
		C	5	Repairing/renovating metal/timber
D:	Groundwork	D	1	Ground investigation/stabilisation/dewatering
		D	2	Excavation/filling
		D	3	Piling
		D	4	Diaphragm walling
		D	5	Underpinning
E:	In situ Concrete/ Large precast concrete	E	1	In situ concrete
		E	2	Formwork

		E 3	Reinforcement
		E 4	In situ concrete sundries
		E 5	Precast concrete large units
		E 6	Composite construction
F:	Masonry	F 1	Brick/block walling
		F 2	Stone walling
		F 3	Masonry accessories
G:	Structural/ Carcassing Metal/timber	G 1	Structural/Carcassing metal
		G 2	Structural/Carcassing timber
		G 3	Metal/Timber decking
H:	Cladding/ Covering	H 1	Glazed cladding/covering
		H 2	Sheet/Board cladding
		H 3	Profiled sheet cladding/covering/siding
		H 4	Profiled panel cladding
		H 5	Slab cladding
		H 6	Slate/tile cladding/covering
		H 7	Malleable sheet coverings/cladding
J:	Water-proofing	J 1	Cementitious coatings
		J 2	Asphalt coatings
		J 3	Liquid applied coatings
		J 4	Felt/flexible sheets
K:	Linings/ Sheathing/ Partitioning	K 1	Rigid sheet sheathing/linings
		K 2	Board/strip sheathing/linings

		K	3	Dry partitions/linings
		K	4	False ceilings/floors
L:	Windows/ Doors/Stairs	L	1	Windows/rooflights/screens/louvres
		L	2	Doors/shutters/hatches
		L	3	Stairs/galleries/balustrades
		L	4	Glazing
M:	Surface Finishes	M	1	Screeds/trowelled flooring
		M	2	Plastered coatings
		M	3	Work related to plastered coatings
		M	4	Rigid tiles
		M	5	Flexible sheet/tile coverings
		M	6	Painting
N:	Furniture/ Equipment	N	1	General purpose fixtures/furnishings/equipment
		N	2	Special purpose fixtures/furnishings/equipment
P:	Building Fabric Sundries	P	1	Sundry proofing/insulation
		P	2	Sundry finishes/fittings
		P	3	Sundry work in connection with engineering services
Q:	Paving/ Planting/ Fencing/Site Furniture	Q	1	Edgings for pavings
		Q	2	Pavings
		Q	3	Site planting
		Q	4	Fencing
		Q	5	Site furniture

R:	Disposal Systems	R	1	Drainage
		R	2	Sewerage
		R	3	Refuse disposal
S:	Piped supply Systems	S	1	Water supply
		S	2	Treated water supply
		S	3	Gas supply
		S	4	Petrol/oil storage
		S	5	Other supply systems
		S	6	Fire fighting - water
		S	7	Fire fighting - gas/foam
T:	Mechanical Heating/ Cooling/ Refrigeration Systems	T	1	Heat source
		T	2	Primary heat distribution
		T	3	Heat distribution/utilisation - water
		T	4	Heat distribution/utilisation - air
		T	5	Heat recovery
		T	6	Central refrigeration/distribution
		T	7	Local cooling/refrigeration
U:	Ventilation/ air conditioning systems	U	1	Ventilation/fume extract
		U	2	Industrial extract
		U	3	Air conditioning - all air
		U	4	Air conditioning - air/water
		U	5	Air conditioning - hybrid
		U	6	Air conditioning - local

	U	7	Other air systems
V: Electrical supply/power/ lighting systems	V	1	Generation/supply/HV distribution
	V	2	General LV distribution/lighting/power
	V	3	Special types of supply/distribution
	V	4	Special lighting
	V	5	Electric heating
	V	6	General/other electrical work
W: Communications/Security/ Control Systems	W	1	Communications - speech/audio
	W	2	Communications - audio/visual
	W	3	Communications - data
	W	4	Security
	W	5	Protection
	W	6	Control
X: Transport Systems	X	1	People/goods
	X	2	Goods/maintenance
	X	3	Documents
Y: Services Reference Specification	Y	1	Pipelines and ancillaries
	Y	2	General pipeline equipment
	Y	3	Air duct lines and ancillaries
	Y	4	General air duct line equipment
	Y	5	Other common mechanical items
	Y	6	Cables and wiring
	Y	7	General electrical equipment

		Y	8	Other common electrical items
		Y	9	Other common mechanical and/or electrical items
Z:	Building Fabric Reference Specification	Z	1	Fabricating
		Z	2	Fixing/jointing
		Z	3	Finishing